Astronomers' Universe

For other titles published in this series, go to
www.springer.com/series/6960

Peter Grego · David Mannion

Galileo and 400 Years of Telescopic Astronomy

 Springer

Peter Grego
PL26 8AS Cornwall
St Dennis, UK
petermoon1@yahoo.co.uk

David Mannion
TN1 2XD Kent
Tunbridge Wells, UK
md.mannion@tiscali.co.uk

ISBN 978-1-4419-5570-8   e-ISBN 978-1-4419-5592-0
DOI 10.1007/978-1-4419-5592-0
Springer New York Dordrecht Heidelberg London

Library of Congress Control Number: 2010933853

Printed on acid-free paper

Springer is part of Springer Science+Business Media (www.springer.com)

Foreword

Galileo Galilei's life and work is one of the great dramas of science, part success and part near-tragedy. His name is honoured, and remembered, by the naming of 2009 – the 400th anniversary of his seminal observations. Galileo's work marks the starting point of an in-depth study of the history of astronomy by David Mannion and Peter Grego – and what a study it is.

The scholarship in this book is excellent. Although a book for school and amateurs – of which there are very many in the world – the depth of treatment is considerable. Nevertheless, all can read it with both pleasure and instruction. The audience should also include professional scientists, indeed anyone with an enquiring mind will find considerable pleasure in its pages.

An unusual but useful feature are the frequent "projects" that the reader is invited to carry out.

The historical chapters form a fine introduction to the eventual description of contemporary astronomy with its own excitements and puzzles.

Galileo Galilei would have been proud of the modern astronomers and also, I think, of Mannion and Grego – who have described so well his discoveries and the exciting science to which they led.

Sir Arnold Wolfendale FRS,
14th Astronomer Royal
July 23, 2009

Preface

For many thousands of years – from the moment that the first thinking human gazed at the skies with curiosity up until the early seventeenth century – people were restricted to viewing the Universe without the use of telescopes. Seasonal cycles, the phases of the Moon, and the motions of the five "wandering stars" were among the first celestial phenomena to be noted. Ever hungry for explanations, humans needed to invent cosmologies to make sense of our place in the Universe. Needless to say, speculation about the cosmos, based partly on observational evidence but mixed with a great deal of conjecture, led to sky lore and saw the incorporation of the heavens into religion. Glorious yet untouchable, the heavens were thought to be an abode of the gods. Throughout the world, in many different human cultures, the heavens were studied in order to divine the plans of the gods, to foretell the future, and to explain great events.

By about the second century BCE, the brilliant work of a number of Greek astronomers such as Aristarchus, Anaxagoras, and Eratosthenes led to the use of mathematics and geometry to attempt to measure the size of the Sun, the Moon, and Earth, and to determine the distances from Earth to the Moon and Sun. The culmination of the astronomical work of the Greeks was brought together by Ptolemy in his great book, the *Almagest*. Fortunately this book was translated by the Arabs in the ninth century CE as Europe passed through the Dark Ages, but the knowledge was passed back and translated into Latin throughout Europe in the eleventh to fourteenth centuries.

It was during the fifteenth and sixteenth centuries that the widely held western view of the Universe as propounded by the Church and based on the *Almagest* was challenged by the likes of Nicholas Copernicus, Johannes Kepler, Tycho Brahe, and Galileo Galilei – great intellects whose own observations forced them to call into question the "unquestionable old knowledge." With his fervent desire to publish the results of his experiments and observations, and his use of the newly invented telescope to make astronomical observations, we have come to regard Galileo as the father of modern science.

A complete change from the long-established geocentric view that anchored Earth firmly to the center of the Universe, to a heliocentric universe with planets and their moons orbiting the Sun, was ushered in at the dawn of the era of the telescope. Later that century Isaac Newton (1643–1727) explained the workings of the Universe with his theories of gravity and kinematics; in the following centuries the distances to the nearest stars were determined, it was realized that our small Solar System is orbiting a vast conglomeration of stars that make up our home Galaxy, and we have discovered that our Galaxy is just one of 100 billion galaxies in an expanding and accelerating universe.

This book is devoted to telling the amazing story of how our knowledge of our Universe was built up during the past 400 years, from the early beginnings of telescopic astronomy through a series of remarkable visual discoveries and to the opening up of the whole of the electromagnetic spectrum and the new astronomies. It is a celebration of the work of generations of astronomers and looks to the exciting future of astronomical research.

While intended to paint a broad picture of the development of telescopic astronomy, a number of intriguing vistas in astronomical history are explored. A complete and exhaustive portrayal of astronomy is, of course, beyond our remit, and if we've neglected to describe certain areas that might have interested the reader, it's not through lack of insight, just lack of space!

Anyway, it is hoped that the excellent work of Galileo and subsequent generations of astronomers will serve as an inspiration to some readers to go and emulate their work: coming soon we have a round of intense solar activity, with sunspots on the Sun reaching maximum due around 2012/2013 and a rare transit of Venus in 2012. Meanwhile, the rest of the Universe is no less amazing to explore and enjoy. Anyone with a small telescope can, for example, marvel at the perpetual waltz of the four Galilean moons of Jupiter, have their eyes opened by the countless stars and deep-sky gems of the Milky Way, and have their retinas tickled by photons millions of years old arriving from the furthermost reaches of the galactic Universe.

Happy observing!

August 2010   Peter Grego and David Mannion

Contents

1. Eyes on the Skies

Our story's central figure, the Italian physicist Galileo Galilei (1564–1642), was intellectually active during the late Renaissance, an exciting and unprecedented period in history. Discoveries in the New World were opening eyes to the idea that our own planet was far vaster and more diverse than anyone had previously dared to suspect. Not only were new discoveries easily disseminated to the educated masses by the printing press, but new ways of thinking about the world were promulgated through pamphlets and books, often couched in the medium of fictional literature.

Galileo Galilei, portrayed by Justus Sustermans in 1636

P. Grego and D. Mannion, *Galileo and 400 Years of Telescopic Astronomy,*
Astronomers' Universe, DOI 10.1007/978-1-4419-5592-0_1,
© Springer Science+Business Media, LLC 2010

Living and working from within some of the Catholic Church's strongest domains – first as Professor of Mathematics at Pisa and later in the Venetian Republic at Padua – Galileo's line of solid scientific reasoning led him to challenge many of the tenets about the Universe that the establishment held dear. Yet, Galileo didn't arrive on the scene amid a vacuum of ideas about how the Universe worked; the trouble was that many of the ideas prevalent at the time were only promulgated through the medium of accepted dogma. Religion has never been very good at shifting its ground once its leaders have proclaimed something to be true, and Galileo's work was truly Earth-shaking.

Some of Galileo's evidence-based notions were far from new – for example, the theory that Earth was a planet in orbit around the Sun had been around since the third century BCE and had famously been postulated by Nicolaus Copernicus (1473–1543) a century before Galileo. However, it was the manner in which Galileo presented his ideas – theories based on sound observations backed up by solid mathematical foundations – that gave the scientist's work such potency. It was with good reason that Albert Einstein (1879–1955) bestowed Galileo with the accolade "the father of modern science." Galileo's work – especially his stance on the correctness of the heliocentric theory – led him to a famous historical clash with the then-powerful Church authorities.

Before looking at the life and work of Galileo, particularly with respect to his work in astronomy and his astonishing telescopic discoveries, it's important to understand how our human view of the Universe and our place within it has been shaped over time.

Pre-Telescopic Astronomical Ideas, Inventions and Discoveries

Although astronomy, the oldest of all natural philosophical pursuits, experienced its dramatic introduction to the telescope just four centuries ago, this age-old science was no stranger to its practitioners' use of mechanical devices for computation and observation. For many centuries astronomical calculations had been aided by wonders such as abacuses and astrolabes, and the heavens had been charted and monitored with quadrants, cross staffs, and a variety of other finely crafted naked-eye instruments.

Regardless of the substantial body of astronomical knowledge that had been acquired since ancient times, though, the telescope proved to be such a potent invention that it changed everything overnight. Not only were the heavens discovered to be more complex and more beautiful than anyone had dared to imagine, it's no overstatement to claim that humanity's perspective upon its own position and status in the Universe changed forever.

Enquiring and observant people throughout the ages have made remarkable efforts to understand the Universe and its workings using nothing but their unaided eyes and raw brain power. So, before exploring the extraordinary development of the telescope, from its humble beginnings through progressively larger and ever-more optically perfect instruments to the incredibly sensitive eyes on the Universe planned for the twenty-first century, let's review some of the important events of pre-telescopic astronomy.

For most of human history, most people lived in small rural settlements and enjoyed truly dark night skies that were untainted by the glare of artificial lighting. Our remote ancestors of millennia gone by could not have failed to have been struck by the sheer grandeur of the heavens. This is a point often unappreciated by most twenty-first century people. More than 3.3 billion people (half the world's population) now reside in cramped towns and sprawling cities. Few occupants of Earth's urban areas appreciate that beyond the orange-tinted artificial sky glow of their night sky lies one of nature's most glorious sights. As urban communities are bathed in permanent light and the night air above glows with countless particles reflecting the glare of street lighting, industry, and commerce, their residents are simply never given the chance to appreciate the darkness. Sadly, few people realize just how magnificent the night skies can be when viewed with the unaided eye.

Our remote ancestors enjoyed incredibly dark skies, where people with average eyesight could easily discern the Milky Way, glowing like a phosphorescent river spanning the skies from horizon to horizon, with dozens of nebulae and star clusters punctuating its course like brighter eddies in the current. Further afield, the nebulous patches of the two close galaxies in Andromeda and Triangulum can be made out indistinctly against the blackness of pristine skies, and southern hemisphere viewers blessed with dark skies delight in the diffuse brightness of the two Magellanic Clouds, our Galaxy's nearest cosmic neighbors.

Against this splendid stellar backdrop – a celestial vault that seemed to revolve around Earth with the passing of the seasons – the Sun made its way unerringly in its annual course along a path known as the ecliptic. The Moon's monthly path among the stars was roughly in line with the Sun's path – hence the name "ecliptic," since every so often the Moon occasionally eclipsed the Sun, and in turn the Moon was eclipsed by the broad shadow of Earth. Eclipse phenomena, along with the regular cycle of lunar phases, gave the Moon an air of profound mystery. Mercury, Venus, Mars, Jupiter, and Saturn – the five brightest planets, all of which have been known since antiquity – each appeared to follow their own individual paths near the ecliptic, but at varying speeds. Their motion among the stars caused them to be known as the "wanderers" (the word "planet" derives from the ancient Greek).

In addition to the more readily visible celestial objects there were many sky phenomena whose appearance was dramatic, unpredictable, and often alarming to our ancestors. From time to time, meteors and fireballs streaked across the night skies; comets sometimes appeared from nowhere to develop into fiery torches with streaming tails; now and again, aurorae lit up the skies in intricate, shifting multicolored displays; occasionally, new stars blazed among the familiar constellations, only to fade into obscurity before long.

The glorious heavens and their remarkable occupants were distant, mysterious, and untouchable. It's hardly surprising that references to celestial objects were deeply embedded in every ancient human culture that we know of, and many ancient religions are full of reverence for sky-related gods. Today it may be tempting to find some amusement at the idea of people fervently praying to the Moon or frightened folk running indoors when an eclipse took place, or an entire nation in panic believing that their world was about to end because of a bright comet that appeared in the skies. But we have the advantage of understanding the nature of these phenomena. Nobody is fond of disorder. People like to have some form of understanding about the world around them. Religious, mythological, and unscientific explanations – however vague, weird or wildly off the mark we might consider them today

– were comforting, and they sat more comfortably with our psyche than any frank acknowledgement of utter ignorance.

One way in which the skies might be better known and understood was to become aware of the motions of its chief players, the Sun and Moon. As well as rising and setting each day, the Sun made a complete circuit around the skies every year, and the highest point it reached above the southern horizon each day appeared to vary with the seasons. Early northern hemisphere cultures based on farming had a great interest in the Sun, and the stones of megalithic structures such as Stonehenge on Salisbury Plain in Wiltshire, England, are chiefly aligned with the cardinal points and the rising point of the midwinter Sun – an important time marking the furthest southern declination of the Sun, after which it begins to climb northwards once more, heralding longer days.

One of the most familiar symbols found in prehistoric sites is the crescent shape, representing the Moon, along with markings indicating the monthly cycle of lunar phases. A lunar calendar comprising 29 markings – one for each day of the lunar month – has been identified in the famous caves at Lascaux in France, and dated to around 15,000 BCE.

That the Moon should be considered important is obvious. The Moon's light was immensely valuable in terms of hunting and nocturnal survival, and the phases of the Moon represented a marker of the passage of time. Carved markings on animal bones found in the Aurignac cave in Haute Garonne, France, clearly show that from about 30,000 years ago the European Aurignacians were avid Moon-chroniclers, for these markings are thought to be meticulous recordings of the passage of lunar cycles. Similar records are found in the art of the western European Azilians (named after a cave near Mas d'Azil in France) made around 4,000 BCE or earlier, and they include narrow crescent shapes that appear to mark New Moons in successive lunations.

One of the earliest depictions of the Moon's physical features (showing the dark patches known as "maria," Latin for "seas") may be a rock carving discovered at a prehistoric tomb in Knowth, County Meath, Ireland, and dating back five millennia; known as Orthostat 47, the carving consists of a set of nestled arcs that bear some resemblance to an impression of the lunar features as viewed at full Moon.

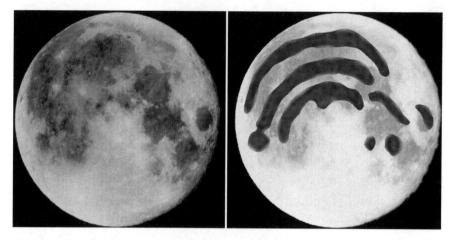

This simple set of lines, taken from an ancient rock carving at a prehistoric site in Ireland, may be the earliest known representation of the Moon's features (Credit: Peter Grego)

It's clear that the skies and its phenomena were initially incorporated into human affairs for practical reasons, and it's not such a great leap from this to astrology – the belief that that the movements of the Sun, Moon, and five planets have an influence on the lives of individual people and the course of world events. A great deal of brain power expended by the earliest civilizations was devoted to watching the skies, noting the movements of the Sun, Moon, and planets, and using this information to predict celestial events, from the movements of Venus to eclipses of the Sun and the Moon. Great civilizations, such as those of ancient Babylon, Egypt, and China, considered it vitally important to observe, record, and predict heavenly phenomena. Astrologer-priests kept a constant vigil on the skies, ostensibly for society's well-being and to keep their rulers informed of any celestial portents that might affect the status quo. Astrology was considered such a precious asset that its use without the ruler's permission was often punishable by execution.

Eastern Skies Under Scrutiny

Ancient China – frequently referred to as the Celestial Empire because of its deep reverence for heavenly events – had an advanced understanding of astronomy. The first Chinese lunar calendars came into practical use around 4,000 years ago. Ancient Chinese astronomers created a catalog of every star visible with the unaided eye, divided the skies into constellations known as "palaces," and referred to the brightest star in each palace as its "emperor star," surrounded by less brilliant "princes." In the fourth century BCE, the astronomer Shih-Shen had cataloged 809 stars and had recorded 122 individual constellations.

Thousands of astronomical phenomena are recorded in Chinese annals dating back many centuries. Over an almost continuous period spanning the sixteenth century BCE to the end of the nineteenth century, Chinese court astronomers were appointed to observe and record changes in the heavens. This legacy of almost 3,500 years' worth of astronomy, in which sunspots, aurorae, comets, lunar and solar eclipses, and planetary conjunctions were recorded in some detail, has provided us with a rich source of reference material.

Astrology played a tremendously important role in ancient China. There were 28 constellations that formed the ancient Chinese zodiac, the band of sky through which the Sun, Moon, and planets appeared to progress. Each of the five planets was designated its own element – Mercury, water; Venus, metal; Mars, fire; Jupiter, wood; Saturn, earth. A person's fate was supposedly determined by the relative position of the five planets, the Moon, Sun, and any comets that happened to be in the sky at the time of that person's birth.

Astrologers to the emperor's court were held in such great esteem that many of them resided within the Imperial Palace itself. Astrologers' advice, based on careful astronomical observations, was highly respected by the emperor, and helped him to make important decisions about the running of the state. Should the emperor's stellar soothsayers make a mistake in their predictions, then dire consequences were bound to befall them; for example, the ancient Chinese *Book of History* records that two court astrologers were executed for having failed to announce a total lunar eclipse in 2,136 BCE.

Relatively infrequent celestial phenomena, such as mutual planetary conjunctions and eclipses, were imagined to be especially potent astrological signs, but the most potent of all heavenly phenomena – those happenings with the power to cause fear and panic – were unpredictable events that took skywatchers completely by surprise. None of these celestial signs was as feared as the arrival of a bright comet. Only a few hundred years ago, comets were almost universally regarded as omens of impending change on Earth, preceding some form of natural catastrophe, famine, pestilence, death of monarch, or war. Ancient astrologers were used to making astronomical observations with the unaided eye (in an era long before the telescope was invented) so the two subjects of astrology and astronomy were once closely intertwined.

Ancient Greek Philosophy

Notwithstanding the penchant our ancestors had for linking events in the skies with happenings on Earth, a substantial amount of astronomical knowledge was built up by ancient civilizations. We know that much of this ancient knowledge has been lost through poor recording combined with cultural changes; for example, much of our knowledge of ancient Greek philosophy comes from texts that were rescued and preserved by Arab scholars during the European Dark Ages. A great library and learning center called the House of Wisdom was instituted in ninth century Baghdad by the Abbasid caliphs Harun al-Rashd and his son al-Ma'mun. Operational until the thirteenth century, the House of Wisdom became the world's greatest center for the study of humanities and the sciences. Ancient Persian, Indian, Greek, and Roman texts – including those of Pythagoras, Plato, Aristotle, Hippocrates, and Euclid – were collected and translated into Arabic. Ancient Greek ideas eventually found their way back into the European arena during the High Middle Ages, as these works were translated into Latin.

Ancient Greek philosophy saw a move away from the age-old acceptance that mysterious supernatural forces under the guise of a pantheon of gods controlled the workings of nature and towards the use of reason and enquiry. A rationalistic approach to philosophy was first espoused by Thales of Miletus (mid-620s–547 BCE), who proposed that it was possible for nature to be understood by humans and that the Universe could be explained in physical terms. Sadly, none of Thales' works survive, but it is thought that he predicted a famous solar eclipse, notable for its occurrence during the Battle of Halys between the Medes and Lydians, an event that appears to have startled the two warring parties into hastily arranging a truce. Using basic geometry Thales estimated that the Sun and Moon – each half a degree across – measured 1/720th of their respective celestial paths.

On the Ball

The popular media often make a joke out of the image of Christopher Columbus's fifteenth-century sailors being afraid that their voyage west was doomed because their little ships would eventually tumble down a stupendous cataract at the edge of the world. While belief in a flat, finite Earth was certainly current in ancient Babylon, Egypt, pre-classical Greece, and China, nobody in Columbus's time seriously thought that our planet was flat. The idea of a spherical Earth was firmly postulated by Plato (427–347 BCE), who imagined that if one could soar high above the clouds our planet would resemble a colored ball similar to those found in an artist's studio. Plato's pupil, Aristotle (384–322 BCE) provided ample evidence to back this up. In *De Caelo* (*On the Heavens*) he rightly claimed that Earth's shape naturally gravitates towards the center, forming a sphere. He explained that the further south one travels, the more southerly or higher the constellations rise in the sky, a phenomenon that could only be explained if Earth had a spherical surface. Moreover, he pointed out that Earth's vast outline can actually be seen during a lunar eclipse, when our planet's curved shadow sweeps across the face of the Moon.

Eudoxus of Cnidus (c. 408–355 BCE) built an observatory and extensively wrote and lectured on astronomy. He invented the astronomical globe – a ball upon which the stars are plotted as though Earth were positioned at its center – a valuable teaching aid that is still used to this day. Eudoxus' *Book of Fixed Stars* contains descriptions of many constellations, including the 12 zodiacal constellations; these star patterns bear a striking resemblance to ancient Babylonian constellations devised several centuries earlier. Eudoxus devised a complete system of the Universe, envisaging Earth at the center of a series of no fewer than 27 transparent crystal spheres upon which were attached the Sun, Moon, individual planets, and stars. Solar, lunar, and more complex planetary movements were explained by the rates at which these spheres rotated. The idea of an Earth centered amid a nest of celestial spheres was to permeate philosophical thought for many centuries.

Aristarchus Takes Center Stage

One notable challenger to the notion of an Earth-centered Universe was Aristarchus of Samos (310–c. 230 BCE), who used observations to support his view that the Sun instead lay at the hub of the Universe. Aristarchus's heliocentric theory of the Universe may possibly have stemmed from his research into determining the Sun's distance using size calculations based on careful observation of the Moon's half phase (dichotomy). With the unaided eye, he judged that at the moment of dichotomy the angle between the Sun, Earth, and Moon was 87°. Since the Moon and Sun have the same apparent angular diameter of half a degree, a simple trig calculation was enough to show that the Sun was 19 times the Moon's distance and was therefore 19 times greater in diameter than the Moon.

With the knowledge that the Sun was nearly seven times Earth's diameter (equating to a volume of around 300 Earths), Aristarchus may have found it a trifle difficult to go along with Eudoxus's notions of an Earth-centered Universe. It turns out that Aristarchus's naked-eye measurements of lunar dichotomy weren't precise enough to give him an accurate figure for the true distance and size of the Sun; he was off by several orders of magnitude. The Sun is actually 400 times further away than the Moon and is 400 times the Moon's diameter (109 times the diameter of Earth and 1,300,000 times Earth's volume).

Eratosthenes Takes Measures

Mathematics and geometry proved invaluable tools in attempting to understand the scale of the Universe. Eratosthenes of Cyrene (276–194 BCE) used observation and trigonometry to derive a number of astronomical distance measurements, among them a remarkably accurate measurement of Earth's circumference in 240 BCE, for which he is best known.

Eratosthenes's logic was simple but powerful. Summer solstice at local noon is the highest point to which the Sun climbs above the southern horizon. He knew that at this date and time the Sun appeared directly overhead at the ancient Egyptian city of Syene (now Aswan), but from his home town of Alexandria, known to be 5,000 stadia (950 km) north of Syene, the solstice noon Sun was 1/50 of a full circle (7° 12') south of the zenith. From this, Eratosthenes deduced that Syene and Alexandria were separated by 1/50 of the total circumference of Earth, which he calculated to be 252,000 stadia – equivalent to 39,690 km, a result within just 1% of the actual value we know today.

Among Eratosthenes's other accomplishments, he developed a system of terrestrial latitude and longitude, drew one of the most advanced maps of the known world, invented the notion of the leap day, accurately calculated the tilt of Earth's axis to its orbit around the Sun, and attempted to work out the Sun's distance from Earth.

Project #1: Measure the Earth's Circumference

If you have a far-off friend on the Internet, you can repeat Eratosthenes's experiment to measure the size of the Earth. For your experiment you need to choose two sites roughly on the same line of longitude, i.e., one site will be due north of the other. Use a vertical gnomon, i.e., a 1 m stick, to cast a shadow on flat ground. The shadow from the gnomon could be measured every 10 min just 40 min before and after noon. The shortest shadow will be at local noon. Using trigonometry you can then measure the altitude of the noon Sun at each site:

Tan θ = height of gnomon/shadow length
Tan θ = 1.00 m/shadow length (in meters)
θ = Tan⁻¹ (1.00 m/shadow length) – find the inverse tan (Tan⁻¹) also known as arc tan button on your scientific calculator.

The distance between sites can be found from the Internet – here are four examples based in the UK and United States.

City 1	City 2	Distance km (miles)
UK		
Cardiff	Edinburgh	497 (309)
Plymouth	Glasgow	609 (378)
United States		
Seattle, WA	San Francisco, CA	1,102 (684)
Pittsburgh PA	Miami, Florida	1,634 (1,015)

The expected difference in altitude of the noon Sun for the four cases would be (assuming the circumference of Earth along a meridian to be 40,008 km) is:

City 1	City 2	Difference in angle (in degrees)
UK		
Cardiff	Edinburgh	4.5
Plymouth	Glasgow	5.5
United States		
Seattle, WA	San Francisco, CA	9.9
Pittsburgh PA	Miami, Florida	14.7

Uncertainties in the measurement of shadow length will produce uncertainties in the final answer, so assuming a simple case of two sites being 7° apart in latitude and the noon Sun having an altitude at the two sites of 40 and 47°, then trying to measure Tan θ you would find the shadow lengths to be 1.19 m and 0.93 m, respectively. A ± 1 cm error in shadow length at both sites would lead to a maximum error of 0.4°, which in turn would lead to error of 2,300 km in Earth's circumference. It is important to make sure the gnomon is vertical – a plumb line using string and a small weight is useful – and that the ground on which the shadow falls is level and horizontal.

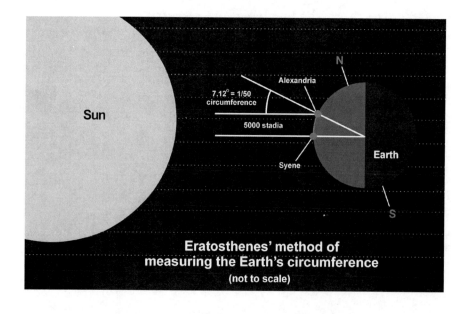

Eratosthenes' method of
measuring the Earth's circumference
(not to scale)

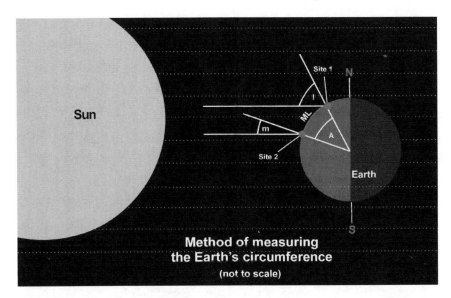

**Method of measuring
the Earth's circumference**
(not to scale)

Hipparchus' Heavens

Of all the great astronomers of antiquity, the Greek astronomer Hipparchus' (190–120 BCE) is most widely considered to be pre-eminent in terms of observational astronomy. In addition to contemporary observations, his work was based upon references to astronomical knowledge and techniques that had been developed in Babylonia long before his time. For example, Hipparchus' possessed an extensive list of ancient Babylonian eclipse observations (records of celestial phenomena originally marked in cuneiform text on clay tablets) spanning many centuries, probably dating back to the eighth century BCE. Using this data, along with contemporary observations and advanced trigonometric tools that he himself had developed, Hipparchus' determined the Moon's motions so accurately that he was able to make reliable predictions of solar and lunar eclipses.

Hipparchus' calculations of the size and distance of the Sun and the Moon were based upon trigonometric reductions of observations. He established, by the use of a naked-eye measuring device called a diopter, that the apparent angular diameter of both the Sun and Moon were half a degree across. Hipparchus' also detected a small variation in the Moon's apparent diameter due to varying distance of the Moon as it orbits Earth. He went on to measure the Earth-Moon distance based upon a solar eclipse of 190 BCE, which happened to have been total at Syene and partial at Alexandria, where one fifth of the Sun (measuring 0.1° across) was visible at maximum. Expressed in radians, this angle enabled the ratio of the distance between Syene and Alexandria to be applied to the distance between Earth and the Moon. Hipparchus' calculated that the Moon lay between 59 and 67 Earth radii distant, the range being based upon assumptions of both the likely minimum distance to the Sun and an infinite distance to the Sun. The average value is actually 60 Earth radii, so his result was remarkably good.

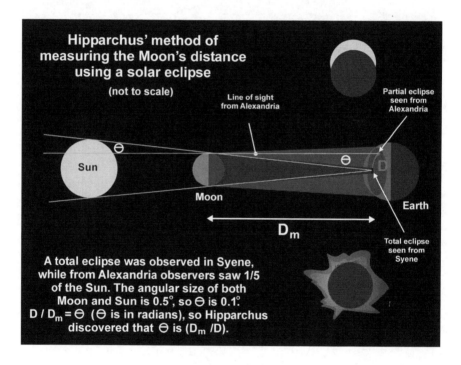

Hipparchus' detailed observations of star positions, made using naked-eye sighting devices including astrolabes and armillary spheres of his own design, enabled him to create the first known stellar catalog around 129 BCE. The catalog featured some 750 stars whose position on the celestial sphere was pinpointed according to his own novel system of celestial longitude and latitude coordinates.

A magnitude scale devised by Hipparchus' denoted each star's apparent brightness; the brightest 20 stars were classed as being of the first magnitude, followed by the next brightest, which were second magnitude, and so on, down to the faintest stars visible, which he classed as sixth magnitude. A similar magnitude scale is used today, although each division between magnitudes corresponds to a precise jump in brightness by a factor of 2.512, and the brightnesses themselves are gauged by photoelectric means. Although no copy of Hipparchus' star catalog exists, it is thought that a representation of it may possibly feature on the intricately sculpted globe borne aloft by the Farnese Atlas, a second century CE marble statue (a Roman copy of an older Greek work) on display at the Naples National Archaeological Museum.

The Farnese Atlas, on display at Naples Archaeological Museum (Courtesy of Gabriel Seah, Wikimedia Creative Commons)

One of Hipparchus' greatest accomplishments was the discovery of the precession of the equinoxes, which he outlines in two books: *On the Displacement of the Solsticial and Equinoctial Points* and *On the Length of the Year*. After measuring the position of the star Spica and comparing his measurements with positional data obtained by Timocharis and Aristillus 150 years earlier, he concluded that Spica had moved by 2° with respect to the point of the autumnal equinox. This, along with discrepancies in the lengths of the tropical year (the time taken for the Sun to return to an equinox) and the sidereal year (the time taken for the Sun to return to a point relative to a star), led Hipparchus' to conclude that the point of the equinoxes was moving along the zodiac from east to west (precessing) at a rate of 46 arcseconds per year. The modern figure for precession is very close, at 50 arcseconds per year, meaning that one complete cycle takes 25,800 years – quite remarkable for such a phenomenon to have been detected in ancient times.

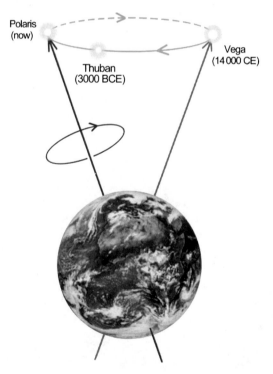

First magnitude Polaris (the Pole Star) is currently the nearest bright star to the north celestial pole. About 5,000 years ago the fainter third magnitude star Thuban was the pole star; in 12,000 years' time, precession will have brought brilliant Vega to the north celestial pole (Credit: Peter Grego)

Loopy but Long-Lasting

Claudius Ptolemaeus (around 83–161 CE) borrowed heavily from earlier philosophers, including the pioneering work of Hipparchus', to produce the *Almagest*. He took care to compile an encyclopedia of ancient Babylonian and Greek knowledge, including the production of a definitive atlas of the stars – no fewer than 1,022 of them, contained within 48 constellations. Expanding on Eudoxus's idea of an Earth-centered Universe, Ptolemaeus explained that the peculiar occasional looping motions (called "retrograde motions") of the planets at some points along their paths were produced when the planets performed smaller circular movements called "epicycles" along their wider circular paths as they orbited the Earth.

While the idea of epicycles seemed to answer a lot of problems and appeared to go some way in explaining the clockworks of the cosmos, careful observations over extended periods of time was later to prove their downfall. These important observations were not to be made for many centuries after Ptolemaeus's time, and the transition from a widespread belief in a geocentric Universe to a heliocentric Universe marks the beginnings of modern astronomy.

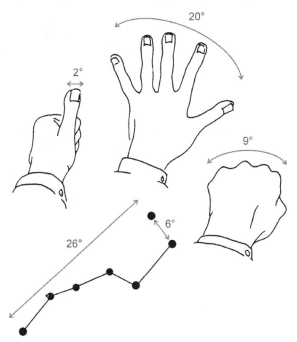

It is possible to make estimations of angles in the sky using just your hand and the unaided eye (Credit: Peter Grego)

Ancient Astronomical Equipment

Nearly everyone owns the most basic of astronomical equipment – brain, eyes, and hands. It's possible to make estimations of angles in the sky using just your hand and naked eye. Here are a few examples. The Moon is half a degree wide – so small that it can be covered with the tip of one's little finger, which is about 1° wide; each day the Moon moves by around 15° towards the east, about the distance between the tips of the outstretched index and little fingers.

Dubhe and Merak, the "pointer" stars in Ursa Major which indicate the direction of the north celestial pole, are separated by 5°, or the width of the three middle fingers. From northern Italy, the midwinter Sun rises to an altitude of around 20°, which equates to the distance between the tip of the thumb and little finger in an outstretched hand; on the same date in Fairbanks, Alaska, the Sun heaves itself to just under 2° above the southern horizon, about the width of a thumb.

A seventeenth-century mariner 'shoots the Sun' using a cross-staff

Carefully constructed instruments will of course enable more accurate measurements to be made. Simple naked-eye devices enabling the measurement of celestial angles have been used since antiquity. More complicated astronomical instruments that permitted calculations to be made in advance included the planisphere and the astrolabe, both of which first appeared in ancient Greece. Consisting of a map of the stars and an overlay that could be rotated to approximate the position of the horizon at any given date and time, the planisphere is an elegant, though rudimentary, device that allows the operator to calculate the rising and setting times of the Sun and stars, and their elevation above (or below) the horizon at any given time. Planispheres are still beloved by amateur astronomers; indeed, most modern astronomical computer programs contain a facility to create a planisphere display.

Astrolabes are a potent combination of the planisphere and a sighting device called a dioptra; thought to have been invented by Hipparchus', astrolabes permitted calculations to be made on the basis of observations, enabling numerous problems in spherical astronomy to be solved. Perhaps the most prolific and proficient exponents of the astrolabe were astronomers of the medieval Islamic world, where they were employed for astronomy, navigation, and surveying, in addition to being put to use for timekeeping for religious purposes.

Instruments to aid naked-eye observations were used extensively in ancient China, as they were in the west. In the first century Lo-hsia-Hung constructed an armillary sphere – a device representing the celestial sphere – upon which were marked 365.25 divisions (for the days of the year) and rings for the celestial equator and the meridian. Lo-hsia-Hung's charming analogy for the Universe likened Earth to the yolk within an eggshell, stating "the Earth moves constantly, but people do not know it; they are as persons in a closed boat; when it proceeds they do not perceive it." In the fifteenth century an observatory was built on the southeastern city wall of ancient Beijing and was equipped with a number of accurately calibrated sighting devices made out of bronze.

A relatively simple naked-eye device called a cross-staff enabled early astronomers to measure the angular distance between individual stars, ascertain the angular heights of stars above the horizon, and track the motion of the Moon and planets against the sky background over a period of time. These simple instruments, whose principles have been known since antiquity, were used extensively by mariners in the late sixteenth and early seventeenth century to take bearings on the angle of the Sun above the horizon at noon (or the altitude of Polaris at night), enabling the ship's latitude to be determined.

The cross-staff consists of a long straight stick upon which is attached a movable vane, known as a "transom," at right angles to it. The observer positions an eye near the bottom of the staff and points it towards the objects whose angular distance from each other is to be measured. The transom is slid up or down the staff so that both objects appear to just touch points (or appear in slits) at either end of the transom. A reading is then taken of the distance of the transom from the eye along the staff.

Since the distance between the two points on the transom is also known, simple trigonometry allows the angle between the two sighted objects to be determined fairly accurately to angular separations of up to 50°, which is the maximum separation of objects that

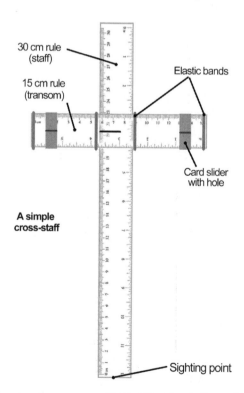

Construction of a simple cross-staff (Credit: Peter Grego)

can comfortably be sighted with the unaided eye. Some of the more elaborate cross-staffs had multiple transoms, or the distances between their transverse points could be adjusted for greater accuracy; calibrated scales on the staff took out the trigonometric work and enabled a direct reading of the angular separation of objects to be made.

Project #2: Make Your Own Naked-Eye Cross-Staff

Materials: 30 cm plastic rule, 2×15 cm thin transparent plastic rules, 4 elastic bands, 2 narrow (1 cm) strips of thin card (length = 4 times the width of the smaller rules). CD marker pen.

Construction: Using a CD marker pen, draw a line lengthways down the center of one of the small rules. Place the small rules on either side of the larger rule at right angles to it, the one you've marked being uppermost. The flat sides of each small rule ought to face each other. Tightly wrap an elastic band around both small rules immediately on either side of the longer rule, so that the two smaller rules are bound together and can be moved up and down relative to the longer rule. The long rule will form the staff, while the smaller rules form the transom. Place a strip of card between the two small rules, one on either side of the main staff, outside of the elastic binding. Fold the card around the small upper rule and create a small tab sticking up from it at the center; tape it firmly so that it maintains its shape. Tightly wrap the other two elastic bands around the ends of the small rules, so that the cardboard tabs can slide along the transom but will stay put whatever position the cross-staff is pointed to. Depending on your preference, punch a hole at the center of each card tab or cut a narrow slit in them. At this scale, this particular cross-staff can be used to measure angles ranging between 15 and 50°.

Observation method: Select two fairly bright stars near to each other in the sky (separated, say, between around one and two outstretched hand widths). Position the eye close to the 0 cm point on the staff and angle the transom so that it's parallel to an imaginary line joining your target stars. Slide the transom so that your target stars can be seen through each hole (or slit). To increase the accuracy of your sighting, try to keep as still as possible during this procedure. Leaning against a wall or bracing your elbows on a suitably solid structure may help. Smaller separations may require you to move the card sliders on the transom closer together. Sighting completed, make a note of the distance of the transom from 0 cm (where you viewed from), and note the separation of the holes (or slits) from each other, in centimeters.

Reduction of observations: You need to work out the angular separation of the stars you've sighted; this entails a simple calculation based on right-angled triangles, using the trigonometric formula

Tan angle = Opposite/Adjacent

"Opposite" (O) is half the distance between the centers of the two holes (or slits). "Adjacent" (A) is the distance between the sighting point on the staff (0 cm) to the center point of the transom.

Let's take an actual example – determining the angular separation between Saturn (magnitude 0.9) and Regulus (magnitude 1.3)

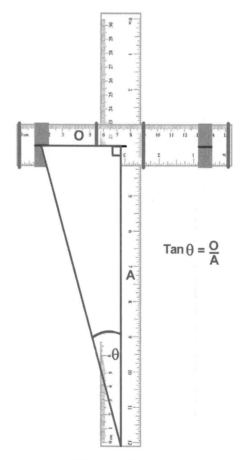

$$\text{Tan } \theta = \frac{O}{A}$$

Trigonometry of a cross-staff (Credit: Peter Grego)

on the morning of January 7, 2009. Our measurements with the cross-staff gave O = 4 cm and A = 23 cm.

Tan angle = Opposite/Adjacent = 4/23 = 0.17

Using tables or a calculator we can convert tangent into the angle.

Tan angle = 0.17 = 11°

Finally, this figure is doubled, giving the observed angular separation of Saturn and Regulus as 21.9°.

By making regular observations it is possible to plot the movements of planets with respect to the stars, revealing phenomena such as retrograde motion and planetary standstills.

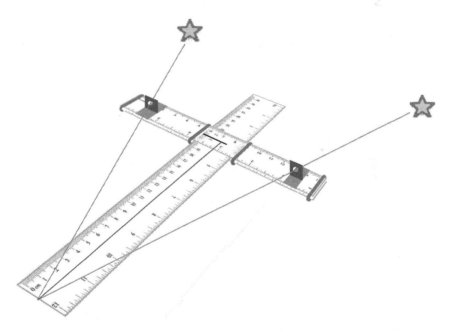

Using a cross-staff to measure angular distances in the sky (Credit: Peter Grego)

The Antikythera Mechanism

Around the year 150 BCE, a cargo ship plying the waters near the little island of Antikythera, half way between mainland Greece and Crete, met with disaster. For some reason unknown to us –

probably the result of a sudden storm – the vessel capsized and sank some 50 m to the bottom of the Kithirai Channel, where it and its cargo remained to gather the usual organic submarine exo-skeleton until it was discovered almost 2,000 years later.

Shipwrecks are not uncommon in this part of the world, as trading between the myriad of islands in the region has been going on since time immemorial. It is thought that this particular ship was laden with loot, en route from the island of Rhodes to the bur-geoning city of Rome. Small items soon began to be recovered from the wreck by sponge divers; among the concreted debris, which included fragments of pottery, sculptures, and coins, several items appeared markedly different to anything that had been previously found at any archaeological site of such antiquity.

Close examination revealed the fragments of a heavily encrusted, corroded, geared device measuring around 33 cm (13 in.) high, 17 cm (6.7 in.) wide, and 9 cm (3.5 in.) thick. Constructed of bronze originally contained within a wooden frame, the device was engraved with a copious text (more than 3,000 characters in length), which appeared to be the device's operating manual. With references to the Sun and Moon, along with the motions of the planets Aphrodite (Venus) and Hermes (Mercury) it was thought that the instrument could have been used to predict various astro-nomical cycles, such as the synodic month (the interval between full Moons) and the metonic cycle (235 lunar months between exact phase repetitions) along with some of the phenomena displayed by the inferior planets. As such, this amazing piece of engineering rep-resents the first portable, programmable computer, demonstrating that the ancient Greeks were far more technologically advanced than they are sometimes given credit for.

Planispheres and astrolabes were used extensively by astrolo-gers in medieval Europe to construct horoscopes. Although we now know that astrology is pseudoscience, without any scientific merit, there was no shortage of eminent practitioners in the west who combined astrology with their more serious astronomical pursuits. For example, Johannes Kepler (1571–1630), brilliant mathemati-cian and originator of the laws of planetary motion, was convinced of the merits of astrology and devised his own system based upon harmonic theory. Some 800 horoscopes formulated by Kepler are still in existence, and certain lucky predictions for the year 1595

The main fragment of the Antikythera mechanism, on display at the National Archaeological Museum of Athens. Despite its condition, the great complexity of the device can clearly be seen (Credit: Marsyas, Wikimedia Commons)

The young Tycho Brahe

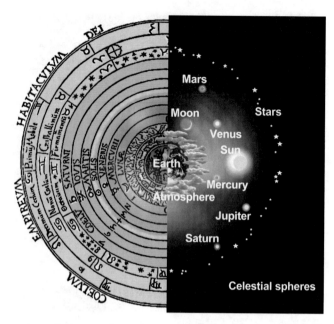

Eudoxus envisaged the Earth at the center of a series of transparent crystal spheres upon which were fastened the Sun, Moon, individual planets and stars (Credit: Peter Grego)

– including foretelling a peasants' revolt, forebodings of incursions by the Ottoman Empire in the east, and predictions of a spell of bitter cold – brought his astrological talents into great renown.

Dogma Defied

Of all ancient Greek philosophical works, those of the polymath Aristotle proved to be the most influential upon Western philosophy. Aristotle's writings on just about every subject that was imaginable in his time – from studies in all the physical sciences to areas of logic and scientific reasoning – were not only consulted by later Greek philosophers but were assimilated into the dogma of the medieval Church, where the ideas held sway from the thirteenth century to the Renaissance.

It was against this deeply ingrained, long-established Church dogma of the truth of Aristotle's views of the nature of the Universe that Renaissance scientists were forced to do battle. The old picture of an Earth-centered Universe – a centrally fixed globe about which the nestled spheres of the heavens revolved – chimed in so comfortably with the Church's idea of humankind's place in the Universe that it proved too comfortable to give up without a fight. The fight was often unfairly balanced in the Church's favor and often took little heed of scientific observations and hard evidence (a situation that may find considerable resonance even today). Slowly, and along a variety of scientific fronts, many of the established dogmatic positions were shown to be false; nevertheless, general acceptance of new ideas often proceeded at a glacial pace.

Sparking a Celestial Revolution

As the Renaissance spread from the sunny shores of the Mediterranean to the bracing climes of the Baltic, enquiring minds across Europe experienced a growing desire to further scientific and astronomical enquiry. Keen, intelligent young minds pondered the cosmos, and the pronouncements of ancient philosophers – long held as solid truths by the all-powerful Church – began to be questioned openly for the first time. Many ancient sources of wisdom were found wanting, and astronomy's biggest paradigm experienced its first assault when Nicolaus Copernicus (1473–1543) suggested that the Sun, not Earth, lay at the center of the Universe.

As we've seen, Copernicus was not the first to suggest that Earth revolved around the Sun. Heliocentric theories can be traced back to ancient Greece, to the philosophers Philolaus and Aristarchus, almost two millennia before Copernicus.

Sometime after 1514 Copernicus was bold enough to produce several copies of a small handwritten pamphlet (but canny enough not to put his name to it) that challenged the very fundamentals of the geocentric view of the Universe, which at the time was accepted by the Church as an undeniable truth. In demoting the status of Earth to a mere planet in orbit around the Sun, he ran the risk of being branded a heretic.

Copernicus was convinced that the Sun, not Earth, lay at the center of the Universe, and that the apparent daily rotation of the heavens is caused by Earth's rotation on its own axis. Of all celestial objects, only the Moon's orbit was centered on Earth. Copernicus went on to explain that the annual circuit of the Sun around the ecliptic is caused by Earth revolving around the Sun, and the apparent periodic retrograde motion of the planets results from the motion of Earth along an orbit inside that of the more slowly orbiting outer planets. His explanation of retrograde motion dispensed with the need to introduce epicyclic planetary motions – circular motions along circular paths, an invention of Ptolemaeus to retain the notion of the perfection of the heavens as exemplified in the circle – and is perhaps the most insightful and original of Copernicus' theoretical points.

Turning his attention to the stars, Copernicus concluded that they are at an immense distance, compared to the distance from Earth to the Sun. His explanation was that the stars appear to remain in exactly the same patterns throughout the year, regardless of the changing point of view on a moving Earth, and therefore the parallax effect on our view of the stars is negligible. Copernicus later laid out his potentially heretical heliocentric views of the Universe in his book *De revolutionibus orbium coelestium (On the Revolutions of the Heavenly Bodies)*, one of the first copies of which was given to him as he lay dying in 1543.

Surprisingly, Copernicus' logic failed to provoke an immediate backlash from the religious authorities. While *De revolutionibus* was attacked several years after its publication by a handful of scholars, the Church's official stance on the heliocentric theory didn't really solidify until 1616, following Galileo's endorsement of the theory and the storm of controversy his work provoked. With an unshakable view that a moving Earth and an immobile Sun idea contradicted the Bible, a Church decree banned *De revolutionibus* along with any other work that promoted Copernicus' heliocentric theory. *De revolutionibus* was eventually removed from the Catholic Church's list of banned publications in 1835, nearly 300 years after its publication.

Hard-Nosed Inquiries

Justly recognized as the founder of modern astronomical observation, Tycho Brahe (1546–1601) was the last (and arguably the greatest) observer of the pre-telescopic era. Of noble Danish birth, his fiery temperament caused him to engage in a sword duel when a young man; Tycho lost part of his nose in the encounter, which he later replaced with a copper prosthesis. His drive to understand the Universe was fueled with tremendous passion.

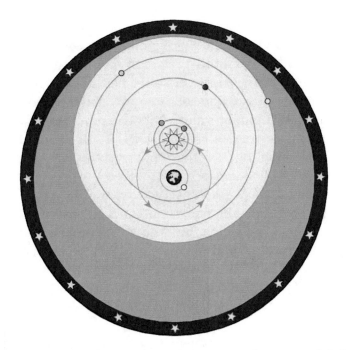

A depiction of the Tychonian geocentric system (not to scale). Both the Moon and the Sun revolve around Earth, while the planets Mercury, Venus, Mars, Jupiter, and Saturn orbit the Sun. In this manner, neither Mercury or Venus appear to stray very far from the Sun, while the outer planets Mars, Jupiter, and Saturn regularly appear at opposition (opposite to the Sun in the sky). The entire Solar System is surrounded by a sphere of fixed stars

Tycho was no fan of Copernicus' heliocentric theory (although he admired its elegant geometry) because he could not discern any stellar parallax owing to Earth's orbit around the Sun. For the Copernican theory to have been correct it would mean that the stars were at an unimaginable distance from Earth, an idea that Tycho found untenable. At the same time, he was unwilling to dispense entirely with the geocentric Ptolemaic system because of the religio-philosophical connotations of removing Earth from the center of the Universe. He therefore set about constructing his own scheme of the Universe, a geo-heliocentric model known as the Tychonic system, in which the Sun and stars orbited Earth while the other planets orbited the Sun. In this scheme, Earth remained at the center of the Universe;

the stars were fixed to a vast, all-encompassing Earth-centered globe, removing the problem of stellar parallax.

Far from being an idle ponderer, Tycho intended to back up his theories by making a multitude of precise measurements of stellar positions and planetary movements. As early as 1563, when aged 17, he had been painfully aware of the inaccuracy of current maps of the heavens, writing: "I've studied all available charts of the planets and stars and none of them match the others. There are just as many measurements and methods as there are astronomers and all of them disagree. What's needed is a long term project with the aim of mapping the heavens conducted from a single location over a period of several years."

Tycho had made a well-deserved reputation for himself as a scientist of great merit; he had established an observatory at Herrevad Abbey (once in Denmark but now in southern Sweden) from where, on the evening of November 11, 1572, he observed a "new star" in the constellation of Cassiopeia. Using self-made naked-eye instruments, Tycho pinpointed the position of his "Stella Nova" and studied it for around a year as it faded from being equal to the brightness of Venus to a dim point of light. Tycho's account of the phenomenon – the first scientifically-studied supernova – was published in Copenhagen in 1573 and was the basis for his subsequent reputation as a first-class astronomer.

Tycho's supernova proved to be an important milestone in the history of astronomy, since nothing like it had been seen before in living memory. Of course, nobody at the time knew that the phenomenon was caused by the explosive death of a star some 7,500 light years away within our own Galaxy (meaning that it had actually exploded around 5900 BCE, when Europe was in the Stone Age and the first signs of civilization had begun to emerge in Mesopotamia and China).

Tycho's dreamed-of long-term project was to be realized in spectacular fashion.

Aware of Tycho's importance in upholding the status of Denmark as a progressive nation that nurtured the advancement of science, King Frederick II granted Tycho a large estate on the island of Hven in Copenhagen Sound (the island now belongs to Sweden), along with generous funding to establish an observatory there. Before long, the astronomer had set about creating the most

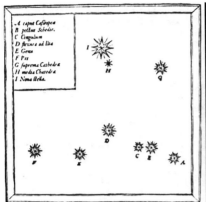

Tycho's published observation of the supernova of 1572, compared with a simulation of what it would have looked like in the sky

More than 400 years after it was observed to explode, the remnant of Tycho's supernova of 1572 is imaged in X-rays and infrared by the Spitzer and Chandra space telescopes. A hot cloud of expanding debris is encapsulated within a spherical outer shock wave outlined by ultra-energetic electrons (Credit: MPIA/NASA)

modern observatory of the day. Between 1576 and 1590, a large castle-styled observatory known as Uraniborg (from its dedication to Urania, the muse of astronomy) grew at the center of the island. Astronomical observations were made using a variety of skillfully

Using a large mural quadrant, Tycho Brahe is pictured at work in his observatory on the island of Hven (From Tycho's *Astronomiae Instauratae Mechanica*, 1598)

fabricated instruments, including a mural quadrant, revolving wooden and steel quadrants, astronomical sextants, and equatorial armillary spheres, many of them featuring novel designs of Tycho's invention.

Eventually, Uraniborg was deemed an unsuitable building for many of his precision instruments, so a nearby area was designated as the main observing site, where a fascinating collection of observatory buildings, known as Stjerneborg (Star Castle), sprang up during 1584. Many of the instruments were contained under cover, largely below ground level in covered waterproof enclosures and domes.

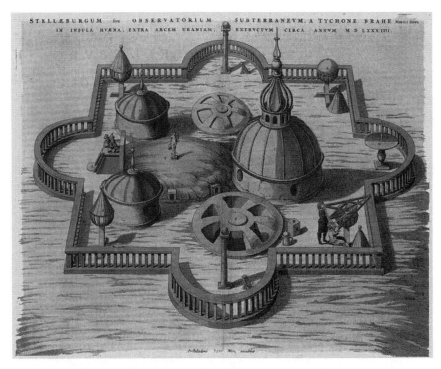

STELLÆBURGUM ſ. OBSERVATORIUM   SUBTERRANEVM, A TYCHONE BRAHE
IN INSULA HVÆNA, EXTRA ARCEM URANIAM,   EXTRVCTVM   CIRCA   ANNVM   M D LXXXIII.

A 1584 illustration of Stjerneborg, Tycho's main observatory

Tycho's meticulous naked eye observations went on to provide a wealth of evidence that not only weighed heavily against the old established Ptolemaic notion of an Earth-centered Universe but also confounded his own Tychonic system. For example, in November 1577 Tycho followed and recorded the path of a bright comet (now designated C/1577 V1) as it made its way across the northern skies. Far from being an object high in Earth's atmosphere, as had been the long-established view of comets since the time of Aristotle, Tycho was certain that the comet lay much further away than the Moon; simultaneous observations of C/1577 V1 with a confederate in Prague, 660 km to the south of Uraniborg, showed that the comet displayed a great deal less parallax than the Moon. Tycho went on to plot the comet's path, placing part of its orbit near that of Venus. He also discovered that the comet's tail always pointed away from the Sun, regardless of the comet's motion through space. These startling discoveries were among numerous thin ends of wedges that were being driven into the once-solid edifice of dogmatic belief in ancient philosophies during the sixteenth and seventeenth centuries. Tycho's 400 year-plus observations of C/1577 V1 – 24 measurements made between November 1577 and January 1578 – have since been used to pinpoint its likely current location, some 300 AU from the Sun, far above the plane of the Solar System.

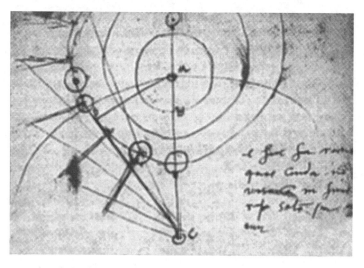

Tycho's notebooks contain an illustration of the path of the comet of 1577 through the Tychonian Solar System

By all accounts, Brahe's growing ideas of his own self-importance tended towards mild despotism; his treatment of the residents of Hven became increasingly arrogant, and it appears that he was prone to delusions of grandeur. After two decades at Uraniborg, Tycho fell out with King Frederick II's successor, King Christian IV, and he left in 1597. The once-proud observatory quickly fell into decline.

Never one to stand still while the Universe moved about him, Tycho secured a prestigious appointment in Prague as Imperial Mathematician to Rudolph II, ruler of the Holy Roman Empire. Continuing his life's work, Tycho began compiling a new set of astronomical tables based on nearly four decades of observation of the stars and the motions of the planets; these were to be called the *Rudolphine Tables* in honor of their imperial sponsor.

Planetary Law-Maker

At Prague, Tycho was assisted in his work on the *Rudolphine Tables* by a gifted young German mathematician named Johannes Kepler (1571–1630), who went on to complete the project following Tycho's death in 1601. Tycho had always clung to the notion of an Earth-centered Universe whose motions would eventually be explained as soon as the right mathematical model was found.

Kepler had none of his master's confidence in the Tychonic theory and used the data contained in Tycho's vast body of work to place the heliocentric theory on firm scientific footing. As a young boy Kepler had marveled at the great comet of 1577 that had proven so difficult to fit into Tycho's scheme of the Universe. Unlike Tycho, Kepler was physically less able to perform the tasks of observational astronomy because of childhood smallpox, which had left him with limb weakness and poor eyesight; instead, his genius lay in the use of mathematics to analyze Tycho's meticulous observations in order to determine a set of fundamental laws about the Universe.

Johannes Kepler, portrayed by an unknown artist in 1610, just after the telescope had first been used by Galileo to reveal astonishing things about the Solar System and the wider Universe

A profoundly religious man, Kepler was convinced that God had created the Universe in accordance with mathematical rules, and that a knowledge of these rules was within human comprehension. Odd then, to think that Kepler should have been well-versed in astrology, the unscientific art of attempting to reconcile the movements of the planets with patterns of events here on Earth. Today, it is strange to think that any rational person could give any credence to astrology, but in Kepler's time the lines between scientific logic and ill-founded superstition were somewhat blurred, particularly when it came to the subjects of astrology and astronomy. Through the practice of astrology as a student at the University of Tübingen in Germany, Kepler first developed an understanding of the apparent motions of the planets; in later life, while at Prague, Kepler's chief role as imperial mathematician to Rudolph II was to provide astrological advice.

In October 1604 – remarkably, just 32 years after Tycho's supernova – another "new star" flared into being. This time, the supernova occurred in the constellation of Ophiuchus and peaked at magnitude –2.5; around 20,000 light years distant; the remnant can still be

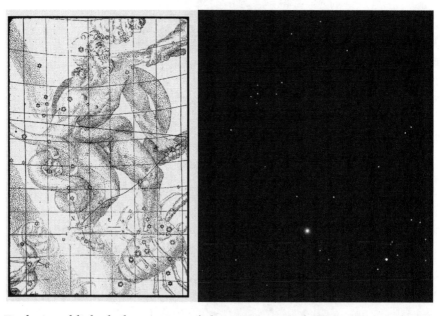

Kepler's published observation of the supernova of 1604, compared with a simulation of what it would have looked like in the sky (Credit: Peter Grego)

observed today. The supernova was also observed by Galileo, who gave lectures on the subject to packed audiences at the University of Padua. Both Kepler and Galileo thought that the supernova affirmed the fact that the heavens could no longer be regarded as changeless and immutable – the new star was undoubtedly in stellar realms because, like other stars, it displayed no measurable parallax. Kepler also interpreted the event astrologically, believing that it represented the beginnings of an important new phase of terrestrial events.

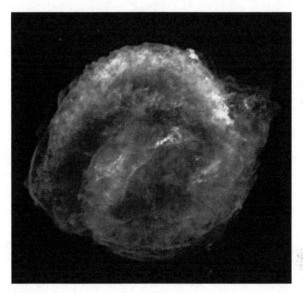

More than 400 years after it was observed to explode, the remnant of Kepler's supernova of 1604 is imaged in X-rays by Chandra space telescope. *Red* colors in the hot cloud of expanding debris represent low-energy X-rays, mainly oxygen; *yellow* shows higher energy X-rays, mostly iron created in the supernova; and *green* shows other elements from the exploded star. *Blue* colors represent the highest energy X-rays, forming a shock front generated by the explosion (Credit: NASA)

While his astrological predictions and horoscopes may have possessed little scientific merit (a quality shared by all astrological forecasts), Kepler's painstaking analysis of Tycho's observations was to produce one of the most important scientific insights of all time. Tycho had originally directed Kepler to investigate Mars's orbit using a mathematical tool known as an equant, helping solve some of the problems observed in planetary motions introduced by the geo-heliocentric Tychonic system. Kepler eventually created a model that agreed with observations to a point, but still produced discrepancies between observation and theory of up to 8 arcmin (almost one quarter the Moon's apparent diameter).

After Tycho's death in 1601, Kepler revisited the problem and attempted to refine the heliocentric Copernican theory by analyzing Tycho's observations of Mars from 1587 to 1595. He dispensed with perfect circles and equants and went to work on the basis that Mars had an ovoid (egg-shaped) path around the Sun. He discovered that the planets move faster when nearest the Sun, and the radius vector (the line connecting a planet to the Sun) sweeps out equal areas in equal times – a concept now known as Kepler's second law of planetary motion.

Kepler's next discovery came with his insight into the shape of Mars's orbit; after years of attempting to reconcile the ovoid, he hit upon the idea of an ellipse – a shape formed by cutting through a cone at an oblique angle. Kepler realized that all planetary orbits must be ellipses of varying degrees – that of Mars being particularly eccentric among the planets – and the Sun is located at one focus of this ellipse. This is known as Kepler's first law of planetary motion. Describing this "eureka" moment, Kepler wrote: "I awoke as from a sleep, and new light broke upon me."

Four years after their discovery, the first two laws of planetary motion featured in his book *Astronomia Nova* (*New Astronomy*), which was published in 1609. The book also contains ideas about gravity (many decades before Isaac Newton's theory of gravitation) and speculation that the Sun's own position in space was far from being stationary. It wasn't until 1619 that Kepler's third law of planetary motion appeared in his book *Harmonices Mundi* (*Harmonies of the World*); it states that the ratio of the length of the semi-major axis of each planet's orbit (cubed), to the time of its orbital period (squared), is identical for all planets. Known as the "harmonic law" this is perhaps better expressed by stating that the square of a planet's orbital period is proportional to the cube of its mean distance from the Sun.

Not only did Kepler's three laws of planetary motion appear to explain the workings of the Solar System, the third law delivered a means by which the scale of the Solar System could be deduced. Kepler's laws enabled Isaac Newton to lay out his theory of gravitation in his *Principia* of 1687, which demonstrates that Keplerian orbits are the most simple of two-body orbits. As we shall see, Newton's Universe is a little more complicated than neatly arranged ellipses.

2. Galileo Magnifico

Not only was Galileo Galilei the greatest of the late Renaissance scientists, but it's no exaggeration to claim that the modern era of physics and astronomy began with his work. Galileo helped revolutionize our thoughts about the way that the world works, opened our eyes to the true majesty of the Universe, and developed the foundations of scientific observation and experimentation to test ideas.

In Galileo's day, the teachings of the Greek philosophers Aristotle and Ptolemy had held sway in the western world for some 1,600 years, helped in no small measure by being thoroughly approved of by the Roman Catholic Church, then the pre-eminent European political power. These ancient imaginings had created a Universe that was made of objects whose nature was perfect and forever unchanging. Aristotle (384–322 BCE) had proposed a Universe with no fewer than 55 crystalline spheres, all rotating about a motionless, central Earth. All the five known planets, the Moon, the Sun, and the stars were supposed to be fastened to these spheres, and they moved around Earth in perfect circles. Galileo's observations could not be reconciled with these ancient geocentric theories of the Universe and instead supported the revolutionary heliocentric Universe of Copernicus, in which the Sun was deemed to be the center of the Solar System and only the Moon revolved around Earth.

Galileo the Scientist

Ever curious about the workings of the world around him, Galileo invented a basic type of thermometer in 1592/1593, consisting of a sealed glass cylinder filled with clear liquid in which floated a number of objects of different densities. Increasing temperature caused the progressively less dense objects to sink to the bottom, and engraved tags attached to the objects could be read to gauge the temperature. Attractive and interesting conversation pieces, working models of

P. Grego and D. Mannion, *Galileo and 400 Years of Telescopic Astronomy*, Astronomers' Universe, DOI 10.1007/978-1-4419-5592-0_2,

these Galilean thermometers can be bought today. Galileo was also famous for his experiments in sound, linking pitch or frequency with the number of notches made in wood by a chisel and the speed at which the chisel was scraped over the wood.

As well as expanding our view of the bigger picture, Galileo is also credited with inventing the compound microscope. In his 1623 book *The Assayer (Il Saggiatore)* Galileo discusses a "telescope modified to see objects very close." Originally called an occhial-ino (small eyeglass), the word "microscope" was bestowed on this device by Galileo's fellow academician Johannes Faber. The first microscopes consisted of two simple lenses; the eye lens was positive, while the objective was negative – the reverse configuration of the optics found in early telescopes. This type of microscope had a very small, dim field of view, since light entering the periphery of the object lens never reaches the eye. A great improvement came with adding a third optical element, the field lens, between the objective and eye lens; this forms an intermediate image from all the available light, which is then magnified by the eye lens, resulting in a brighter image with a larger field of view.

A late seventeenth-century modified Galilean microscope, constructed from cardboard, leather, wood, and iron, used three bi-convex lenses – an objective, a field lens, and an eyepiece (Credit: Peter Grego)

Getting the Ball Rolling

Among his most noteworthy scientific achievements, Galileo is credited with being the first to observe the isochronism of a swinging pendulum, wrote treatises on centers of gravity and hydrostatic balances, experimented with rolling balls down inclined planes, and understood the parabolic nature of the paths of projectiles. In physics, Galileo is famous for introducing the concept of inertia, formulating laws of acceleration, and testing his hypotheses by experiment.

Galileo attempted to measure the rate of acceleration due to gravity by rolling balls down an inclined plane. He found that the distance traveled by a ball was proportional to the time squared. He also is said to have experimented by simultaneously dropping balls of different mass from the Tower of Pisa. Aristotle had thought that a ball with a larger mass would fall more quickly than a ball with a lighter mass – a cannonball would thud to the ground more quickly than an inflated pig's bladder of the same shape and size. However, Galileo's experimental result was that the acceleration due to gravity – and thus the duration of fall – was practically the same for balls of different mass. This was dramatically proved in 1971 by *Apollo 15* Commander David Scott during one of his Moon walks when he dropped a 1.6 kg geologist's hammer and a falcon's feather. Without any air resistance to impede their descent, both objects fell to the lunar surface at exactly the same rate.

Galileo's ideas of inertia and the concept that a force is required to change the velocity of a body were forerunners to the Laws of Motion developed by Sir Isaac Newton in 1687. Galileo developed ideas now known as Galilean relativity, in which he surmised that an observer traveling below deck in a ship (without the benefit of a porthole) could not through experiment determine if the ship was moving. You may have had a similar sort of experience while sitting on a train in a station when the train alongside you has started to move. Can you tell whether it's you or the train next to you which is moving, or both? Now close your eyes – can you tell that Earth is moving at 30 km/s around the Sun? Do you sense that the Sun and all its attendant planets are moving at 20 km/s around our Galaxy?

Project #3: Measuring Acceleration Due to Gravity

Aim: Show that Distance (D) travelled down an incline plane is proportional to time squared (t²).

Directions: Take a very smooth plank of wood and set it at a slight angle to the horizontal – a difference in height of 2 cm over a length of 1 m would be a good start. Mark off distances down the plane every 20 cm. Now time a small ball rolling down the plank, measuring the time taken to reach 0.2 m, 0.4 m, and so on up to 1.0 m by taking an average of three timings. Plot a graph of time squared (t² in s²) against distance (D in m) traveled down the slope. A straight line graph passing through the origin will show that there is a proportional relationship between distance traveled down the incline and the time taken squared. If you raise the plank for further measurements you will find that the acceleration is greater.

Tip: Use a video camera or your mobile phone's video to record the experiment. That way you can replay the experiment and time it more accurately.

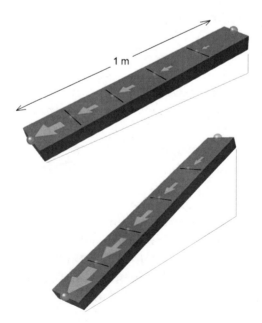

Rolling a ball down an inclined plane (Credit: Peter Grego)

Galileo's Telescope

Credit for the invention of the telescope is usually attributed to the Dutch-German lens maker Hans Lippershey (1570–1619) of Middelburg, Zeeland, in the Netherlands. A traditional story tells us that youngsters (perhaps his own children) playing in his workshop stumbled upon the fact that the combination of a negative (concave) and a positive (convex) lens will magnify a distant image, provided that the negative lens is held near the eye and the lenses are firmly held at the right distance from each other; just why children would be allowed to play in his workshop full of delicate and expensive glass items is not explained, and of course the story is utterly unverifiable.

Regardless of whether this fortunate discovery was made by accident or by careful experiment, Lippershey presented his invention – a

Hans Lippershey, inventor of the telescope, portrayed by Jan van Meurs in 1655

device he called a kijker (a "looker," which magnified just three times) – to the Dutch government in October 1608, with the intention of obtaining a patent, stating that such an instrument would have enormous military potential. However, it was thought that there was little chance of successfully keeping the invention a secret or preventing others from making their own telescopes, and the patent was declined. Nevertheless, Lippershey was well rewarded for his design, and he went on to make several binocular telescopes for the government.

Two other Dutch opticians later claimed to have come up with the idea of the telescope prior to Lippershey – Jacob Metius (1571–1628) of Alkmaar in the northern Netherlands, who actually filed his patent application just a few weeks after Lippershey, and the notorious counterfeiter Sacharias Jansen of Lippershey's home town of Middelburg, who claimed to have made a telescope as early as 1604. However, Lippershey's patent application represents the earliest known documentation concerning an actual telescope, so the credit rightly remains with Lippershey.

News of the marvelous invention quickly spread around Europe, and first reached the ears of Galileo in May 1609. By July of that year Galileo, working without any special knowledge of Lippershey's invention, had figured out what kinds of lenses were required and built his first refracting telescope. Consisting of a pair of lenses in an adjustable tube – an objective lens to collect and focus the light and an ocular lens to magnify the focused image – Galileo's first telescope only magnified about three times. Galileo wrote that his telescope consisted of:

> ...a tube, at first of lead, in the ends of which I fitted two glass lenses, both plane on one side, but on the other side spherically convex, and the other concave. Then applying my eye to the concave lens I saw objects satisfactorily large and near, for they appeared one-third of the distance off and nine times larger than when they are seen with the natural eye alone.

Galileo's second telescope, constructed a few weeks later, was of the same configuration but had a magnification of 10×. He cleverly found an opportunity to show it off to a group of dignitaries who used it to spot distant objects from the bell tower of San Marco in Venice. They were so impressed with what they saw (and no doubt impressed with its potential military applications) that Galileo's salary was doubled, and he was given a lifetime appointment to the Padua Chair of Mathematics.

Galileo's first telescopes were low-powered refractors that used off-the-shelf spectacle lenses. By November 1609 he had constructed a telescope with a magnification of 20×, grinding the lenses himself to his own specifications in order to produce an instrument with a higher magnification. It consisted of a 37 mm plano-convex objective lens (outward curving on one side and flat on the other) with a focal length of 980 mm; a smaller plano-concave lens (inward curving on one side and flat on the other) was used as an eyepiece to magnify the image. The tube was of a wooden barrel-type construction, made of long strips of wood glued together, and consisted of two parts – the main tube, to which the objective was fixed, and a smaller drawtube nestling inside it, which housed the eye lens. In the process of lens selection he assessed the quality of dozens of lenses, taking great care to use only the best of these. As a result, Galileo's telescope was optically far superior to instruments made by most other telescope makers in Europe at the time, enabling him to make a number of remarkable (and quite unanticipated) discoveries about our Solar System.

One of Galileo's original telescopes (Credit: Peter Grego)

Project #4: Make Your Own Galilean Telescope

You can experience the same kind of views that Galileo had of the Universe by constructing your own Galilean-type telescope – a simple refractor that uses two small single lenses. Cardboard or PVC tubing makes a good, inexpensive telescope tube. One lens, the objective, is fixed at the far end of the telescope tube and collects and focuses incoming light; the other lens, the eyepiece, magnifies the image and is mounted in a smaller cardboard tube that is capable of being slid backwards and forwards to achieve a sharp looking image. The objective lens needs to be positive – that is, having either a bi-convex surface or a plano-convex surface; the eye lens must be negative, a bi-concave or plano-concave shape. Galileo's telescope had a plano-convex objective and a plano-concave eye lens. (Just one surface on each lens had been optically shaped in the workshop, hence making it easier to produce.)

The overall length of the finished telescope, and the magnification that it delivers, will depend on the focal lengths of both lenses. The magnification of any telescope can be worked out by dividing the focal length of the objective lens (or main mirror, if it's a reflector) by the focal length of the eyepiece. So, a Galilean telescope with a 30 cm focal length objective combined with an eye lens of 5 cm focal length will magnify six times (30 divided by 5); the telescope will be about 35 cm long.

All other things being equal, the shorter the telescope, the more distorted the view will appear, because singlet objective lenses with short focal lengths produce greater degrees of aberration. Chromatic aberration, where false colors appear in an image, is caused when a lens fails to bring all wavelengths of visible light to a single focal point. Another form of aberration occurs because a lens with a spherical surface focuses light to different points along the optical axis (a line running from the center of the lens and at right angles to it), producing distortion called spherical aberration.

For the purposes of this book (the authors won't recommend anything unless they've tried it for themselves), a very small Galilean telescope was made using a 2.5 cm plano-convex objective with a focal length of 11 cm and a 2 cm plano-concave eye lens with a 5 cm focal length. An inexpensive 10×25 binocular provided the positive objective lens, while the negative eye lens came from an old unused eyepiece. Both lenses simply unscrewed from their original housings without any problems. If you don't have any spare optical gear lying around waiting to be temporarily assimilated into this retro-project, you can obtain inexpensive lenses from a wide variety of sources. In the United States, the long-established Edmund Scientific has a website that sells a huge selection of optical components, including lens kits that are perfectly suited to projects like this. Anchor Optics sells surplus lenses with diameters and focal lengths suited to making a good replica Galilean telescope, and many more can be found by doing

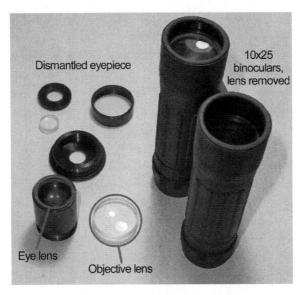

A temporarily dismantled binocular and telescope eyepiece provided the two lenses for the Galilean telescope (Credit: Peter Grego)

a little googling. Unless you're making a keepsake replica from the highest quality materials that match those used by Galileo, the project shouldn't cost more than $15–$20 in total.

Construction was fairly straightforward. First, thick gauge grip tape was added around the edges of both lenses. Next, two tubes were made from thin card – one for the objective (the main tube) and a smaller one for the eyepiece drawtube; the card was simply rolled around the lenses and secured in place on the outer edge with Scotch tape. Obviously, the length and width of the tubing will vary with the focal length of the lenses chosen; our telescope tubing was made from a single sheet of A4-sized card. An attractive design based on the original colors of Galileo's telescope was used as the print for the exterior of the card; note that to improve the quality of the telescopic view, the interior of the tubing should ideally be colored matt black. Enough card was rolled around the eye lens to create a snug fit between it and the main tube, so that

Grip tape houses the objective and eye lenses (Credit: Peter Grego)

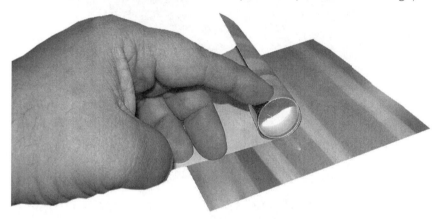

The lenses are rolled in thin card to create the main tube and drawtube (Credit: Peter Grego)

The finished telescope – small and neat (Credit: Peter Grego)

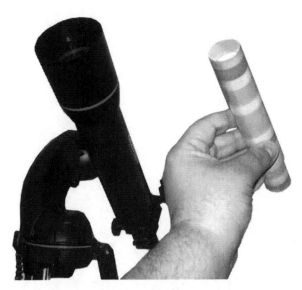

The little telescope compared with a modern computer-controlled achromatic refractor – both are fun to use (Credit: Peter Grego)

the eyepiece is able to be moved in and out over a short distance to achieve focus.

Galilean telescopes give a right way up image, but they have a rather small field of view compared with a typical amateur astronomer's eyepiece. They can be held in the hand quite comfortably at low magnifications, but some form of steady support is required if your telescope magnifies more than 10×. Our little telescope magnified a little over 2× and was easy to use "in the field"; through it the larger lunar craters could easily be distinguished.

The Starry Messenger

Astonished at the celestial sights that he viewed through his new telescope in late 1609 and early 1610, Galileo worked at a lightning pace to publicize his new discoveries. In March 1610 he published his startling observations in *Sidereus Nuncius* (*The Starry Messenger*), a little book that made a big impact. In addition to revelations about the Solar System – notably discussions about the Moon's cratered face and the moons of Jupiter – the book included observations of the Milky Way, stars in Orion, and the bright Pleiades and Praesepe open star clusters.

SIDEREVS
NVNCIVS
MAGNA, LONGEQVE ADMIRABILIA
Spectacula pandens, suspiciendaque proponens
vnicuique, præsertim verò

PHILOSOPHIS, atq; ASTRONOMIS, quæ à

GALILEO GALILEO
PATRITIO FLORENTINO
Patauini Gymnasij Publico Mathematico

PERSPICILLI
Nuper à se reperti beneficio sunt obseruata in LVNÆ FACIE, FIXIS IN-
NVMERIS, LACTEO CIRCVLO, STELLIS NEBVLOSIS,
Apprime verò in
QVATVOR PLANETIS
Circa IOVIS Stellam disparibus interuallis, atque periodis, celeri-
tate mirabili circumuolutis; quos, nemini in hanc vsque
diem cognitos, nouissimè Author depræ-
hendit primus; atque

MEDICEA SIDERA
NVNCVPANDOS DECREVIT.

VENETIIS, Apud Thomam Baglionum. M DC X.
Superiorum Permissu, & Priuilegio.

The title page of Galileo's groundbreaking book *Sidereus Nuncius*. It reads:
"The Starry Messenger: unfolding great and highly admirable sights. Presenting to the gaze of everyone, especially philosophers and astronomers,
those things observed by Galileo Galilei, patrician of Florence, Public
Mathematician of the University of Padua, with the aid of a telescope
which he has recently devised: the face of the Moon, innumerable fixed
stars, the Milky Way, cloud-like stars, and especially concerning four
planets revolving around Jupiter with unequal intervals and periods, with
wonderful swiftness, which, known to no-one up to this day, the Author
most recently discovered for the first time, and determined to name the
Medicean Stars"

Galileo's telescopic discoveries, and his scientific assessment of what he had observed, were soon to shatter the accepted view that all bodies in the Universe were perfect. His observations of mountain ranges and craters on the Moon and spots on the Sun were in direct opposition to this idea. Moreover, Galileo's observations clearly showed that Earth wasn't located at the center of the Universe. Planets exhibited discs, the planet Venus showed phases (just as the Moon does), and four star-like points (now known as the Galilean moons) were obviously in orbit around the planet Jupiter. None of these facts could be explained using the old geocentric Ptolemaic system.

It's worth taking a more detailed look at Galileo's telescopic observations and discoveries, beginning with the Moon (the first astronomical object to be scrutinized by the telescope), moving to the Sun and planets out to the stars themselves, examining what scientific deductions Galileo made on the basis of his observations. All of Galileo's observations can be replicated using a small telescope, so we have included a number of interesting observational projects along these lines. We've also compiled a useful list of future dates suitable for observing the Moon in identical phases to those scrutinized by Galileo, along with a listing of dates when Jupiter's satellites will be lined up in the same sequence as when Galileo observed them. Given that we have the benefit of four centuries of astronomical discovery between us and Galileo, the informed modern observer using modest equipment is fortunate to be able to understand various astronomical phenomena with somewhat more perspicacity than Galileo.

The Moon

Measured against the distant stars, the Moon appears to orbit Earth in 27.3 days (known as the sidereal month), but with respect to the Sun it takes 29.5 days (a synodic month). Why the discrepancy? Well, Earth is in orbit around the Sun, so for the Moon to be in the same place in its orbit relative to the Sun it takes an extra 2.2 days. The Sun always lights up one half of the Moon's surface, but an observer on Earth sees different phases, waxing from a thin crescent, through first quarter, gibbous to full, when the Sun, Earth, and Moon are

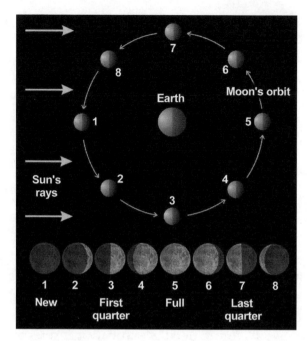

The Moon goes through a sequence of phases every 29.5 days (Credit: Peter Grego)

in a line; after full, the Moon goes through the same sequence in reverse, waning through gibbous, through last quarter to new, when the Moon's day side is turned away from Earth and can't be seen.

Harriot's Lunar First

Galileo was keen to record impressions of what he could see through the telescope eyepiece by making observational sketches. It's clear from his notebooks that Galileo had a well-developed artistic talent – his pen and inkwash depictions of the Moon are delightful examples of his fine draftsmanship.

Galileo's first lunar drawing was made on November 30, 1609, but he was not the first person to sketch a telescopic view of the Moon. That honor goes to a remarkable Englishman, Thomas Harriot (c.1520–1621), who made a series of sketches of the Moon as seen through his telescope, beginning in the summer of 1609, several months before Galileo.

Although Harriot led a remarkable life and had friends in some of the highest places in Elizabethan society, we know little about the man himself; from an age when portraits abound, no depictions of him are known. A gifted mathematician with a keen, enquiring mind, Harriot accompanied Sir Walter Raleigh on a voyage to North America in 1585–1586; he spent a while on Roanoke Island off the coast of North Carolina and learned the Algonquian language. His subsequent account of the voyage, published in 1588 – with its positive account of the native population and talk of the riches this new land might yield – proved influential in encouraging English exploration of the New World.

Harriot became interested in astronomy in 1607 when, from the gardens of Sion House in London, he observed Halley's Comet glide through the northern skies. This was not the comet's most conspicuous apparition; Kepler described it as being a first magnitude star, and its faint tail was just 7° in length. Those of us who strained to catch a glimpse of Halley's Comet in 1985–1986 will think that this was a pretty good showing, nevertheless! Kepler (unaware that Halley's Comet was locked in a 76-year-long orbit around the Sun) calculated that this particular comet moved through the Solar System along a straight path. Following the appearance of three bright comets in 1618, Galileo argued in his *Discourse on Comets* (1619) that these objects were simply optical phenomena – light reflected from clouds at high altitude, rather than distant bodies moving around the Solar System.

In 1609 Harriot purchased a telescope, calling the device a "Dutch trunke" (it was either imported from Holland or made by John Tooke, Harriot's instrument maker). His initial observational lunar sketch – the world's first depiction of another celestial object viewed through a telescope – was a simple pen and ink line drawing of the wide waxing crescent Moon on the evening of August 5, 1609, viewed at a magnification of 6. Note that this is the corrected Gregorian calendar date; prior to the Gregorian calendar's introduction in Britain on September 14, 1752, the Julian calendar was used, and Harriot actually dated his observation July 26, 1609. Although the crescent phase appears exaggerated in this sketch, the horns of the Moon curving in too strongly, the roughness of the lunar terminator caused by craters and mountains along the

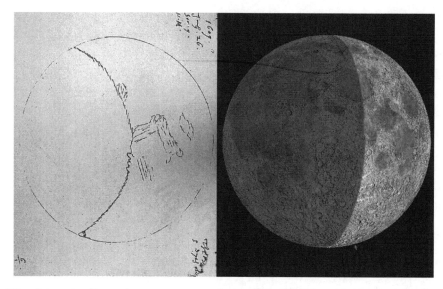

The historic first close-up view of another world – Thomas Harriot's lunar observation of August 5, 1609 (*left*), compared with a *Virtual Moon Atlas* computer simulation of the same view (Credit: Right photo by Peter Grego)

line of sunrise is depicted well, along with coarse shading to represent the eastern maria, with Mare Serenitatis bisected by the terminator in the north.

Harriot went on to make a number of observational lunar drawings during 1609 and 1610, his accuracy and drawing technique improving as time went by. In 1610 he produced a map of the Moon – something that Galileo never did – a detailed and surprisingly accurate line drawing annotated with a set of observational reference points for various prominent lunar surface features. Sadly, Harriot never published his work, but he did encourage some of his close friends to observe the Moon for themselves. Harriot's pupil, Sir William Lower (1570–1615) was viewing the Moon through a telescope supplied by Harriot several months before Galileo's first observations. In a letter written to Harriot, Lower's description of the Moon's surface is remarkable:

> According as you wished I have observed the moone in all his changes. In the new manifestlie I discover the earthshine a

little before the dichotomie; that spot which represents unto me the man in the moone (but without a head) is first to be seene. A little after, neare the brimme of the gibbous parts towards the upper corner appeare luminous parts like stares; much brighter than the rest; and the whole brimme along looks like unto the description of coasts in the Dutch books of voyages. In the full she appeares like a tart that my cooke made me last weeke; here a vaine of bright stuffe, and there of darke, and so confusedlie all over. I must confesse I can see none of this without my cylinder.

Galileo's Moon and Earthshine

With its white clouds and polar ice caps, Earth's albedo (its reflectivity) is very high. Some 39% of incident sunlight is reflected into space by our planet. Some of this light hits the Moon and illuminates portions of the lunar surface that aren't in direct sunlight. Surprisingly, the Moon itself is not a very good reflector. Only around 7% of incident light is reflected by the lunar surface back into space. Those parts of the Moon illuminated by light from Earth appear to glow faintly, a phenomenon known as Earthshine, and this glow is visible with the naked eye when the Moon is in a waxing crescent phase between 1 and 5 days old, and again when it's in a waning crescent between 23 and 27 days old.

Viewed from the Moon, the full Earth is 2° across – four times the diameter of the Moon as seen from Earth – and therefore has an area some 16 times that of the Moon. Since Earth's albedo is around five times that of the Moon, the full Earth seen from the Moon is some 80 times brighter than the full Moon seen from Earth.

Leonardo da Vinci (1452–1519) correctly attributed the Earthshine to light reflected by the Earth back in 1506. The phenomenon was observed by Galileo and correctly interpreted in *Sidereus Nuncius*, where he stated: "The Earth, in fair exchange, pays back to the Moon an illumination similar to that which it receives from her throughout nearly all the darkest gloom of night."

Earthshine adorns the young crescent Moon (Credit: Peter Grego)

A New World

When he first turned his telescope to the Moon, Galileo was amazed to discover that subtly mottled disk of old transformed into a rugged world that looked very much like parts of Earth. Large dark areas that Galileo calls "great spots" have been known since time immemorial; these features, known as the Moon's "maria" (Latin for "seas"), were observed to be relatively smooth, dark plains. In addition to the maria, small dark spots seen within the illuminated parts of the Moon appeared to change over time. According to Galileo, these spots "always agree in having their blackened parts directed toward the Sun, while on the side opposite the Sun they are crowned with bright contours." He concluded that they could be nothing but large lunar valleys. We now call these often deep-looking circular features craters, a word originating from the Latin for "mixing-bowl."

Noting that bright points of light often appear in the dark portion of the Moon beyond the terminator (the division between the illuminated and unilluminated lunar hemispheres), Galileo concludes that they are mountain summits catching the Sun. He wrote: "...on the Earth, are not the highest peaks of the mountains illuminated by the Sun's rays while the plains remain in shadow?" Galileo was also bold enough to make one specific comparison between Earth and the Moon. He likens one particular feature – thought to be the large crater now known as Albategnius – to what the district of Bohemia (in the Czech Republic) would look like if viewed from above.

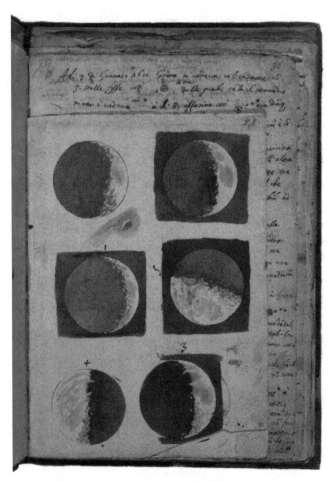

This page from Galileo's original notebook shows a series of lunar observational sketches, each made at the telescope eyepiece. Unfortunately none of these drawings is accompanied by a date and time, so some careful investigation has been required to determine when they were made

Project #5: Observing Opportunity – View Galileo's Moon

Using steadily held binoculars or a small telescope it's possible to view the Moon at the same phase and libration that Galileo saw and depicted in his seven inkwash depictions and the four coarser engravings published in *Sidereus Nuncius*. Galileo failed to assign dates and times to any of these observations, so some detective work has been necessary to make an educated guess as to when they were made. Ewen Whitaker, a highly respected astronomer specializing in lunar studies, calculated the following dates and times for each of Galileo's observations:

Drawing ID	Colongitude[a]	Date	Time (UT)
F1	321.2	30 Nov 1609	15:00
F2	322.2	30 Nov 1609	17:00
E1	323.2	30 Nov 1609	19:00
F3	334.1	01 Dec 1609	16:30
F4	346.0	02 Dec 1609	16:00
E2	358.2	03 Dec 1609	16:00
E3	162.1	17 Dec 1609	04:00
F5	162.1	17 Dec 1609	04:00
E4	174.2	18 Dec 1609	04:00
F6	175.2	18 Dec 1609	06:00
F7	204.2	19 Jan 1610	05:50

[a]Colongitude is a measurement denoting the position of the Moon's terminator

The following list gives suitable dates corresponding to Galileo's lunar observations, each with a matching colongitude and the Moon displaying a similar libration (although the latter is not exact in terms of position and extent).

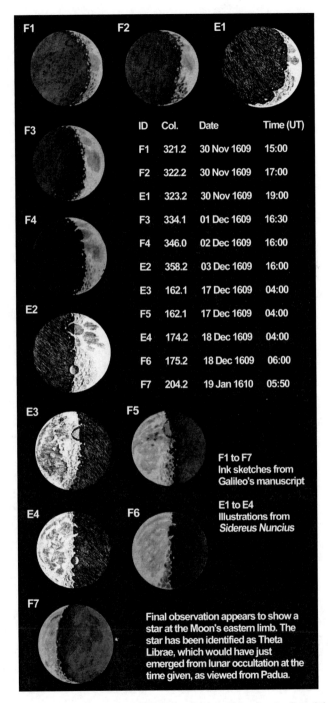

ID	Col.	Date	Time (UT)
F1	321.2	30 Nov 1609	15:00
F2	322.2	30 Nov 1609	17:00
E1	323.2	30 Nov 1609	19:00
F3	334.1	01 Dec 1609	16:30
F4	346.0	02 Dec 1609	16:00
E2	358.2	03 Dec 1609	16:00
E3	162.1	17 Dec 1609	04:00
F5	162.1	17 Dec 1609	04:00
E4	174.2	18 Dec 1609	04:00
F6	175.2	18 Dec 1609	06:00
F7	204.2	19 Jan 1610	05:50

F1 to F7
Ink sketches from
Galileo's manuscript

E1 to E4
Illustrations from
Sidereus Nuncius

Final observation appears to show a star at the Moon's eastern limb. The star has been identified as Theta Librae, which would have just emerged from lunar occultation at the time given, as viewed from Padua.

Possible dates and times (calculated by Ewen Whitaker) of Galileo's seven inkwash sketches from his manuscript and the four lunar illustrations featured in *Sidereus Nuncius* (Credit: Peter Grego)

Drawing ID	Date	Time (UT)
F1	7 May 2011	05:45*
	25 Apr 2012	15:00*
	9 Sep 2013	08:15*
	22 Dec 2017	13:00*
F2	7 May 2011	07:45*
	25 Apr 2012	17:00
	9 Sep 2013	10:15*
	22 Dec 2017	15:00*
E1	7 May 2011	09:45*
	25 Apr 2012	19:00
	9 Sep 2013	12:15*
	9 Sep 2013	12:15*
	22 Dec 2017	17:00
F3	8 May 2011	07:00*
	26 Apr 2012	16:15
	26 May 2012	03:45*
	24 Jun 2012	15:00*
	14 Jun 2013	00:45*
	13 Jun 2013	11:30*
	11 Aug 2013	22:30*
	1 Feb 2017	20:45
	23 Dec 2017	14:15*
F4	9 May 2011	06:15*
	27 Apr 2012	15:45*
	27 May 2012	03:15*
	25 Jun 2012	14:30*
	15 Jun 2013	00:15*
	14 Jul 2013	11:00*
	12 Aug 2013	21:45
	2 Feb 2017	20:15
	24 Dec 2017	13:45*
E2	10 Apr 2011	18:45
	10 May 2011	06:15*
	28 Apr 2012	15:45*
	28 May 2012	03:30*
	26 Jun 2012	14:30*
	17 May 2013	13:00*
	16 Jun 2013	00:15
	15 Jul 2013	11:00*

(continued)

(continued)

Drawing ID	Date	Time (UT)
	13 Aug 2013	21:30
	3 Feb 2017	20:15
	25 Dec 2017	13:45*
E3	24 Apr 2011	05:00
	23 May 2011	16:45*
	12 Apr 2012	14:15*
	12 May 2012	02:00
	10 Jul 2012	00:30
	30 May 2013	23:15*
	8 Jan 2018	01:45
F5	24 Apr 2011	05:00
	23 May 2011	16:45*
	12 Apr 2012	14:15*
	12 May 2012	02:00
	10 Jul 2012	00:30
	30 May 2013	23:15*
	8 Jan 2018	01:45
E4	25 Apr 2011	05:00
	24 May 2011	16:45*
	13 Apr 2012	14:00*
	13 May 2012	02:00*
	11 Jun 2012	13:30*
	11 Jul 2012	00:30
	2 May 2013	11:30*
	31 May 2013	23:00*
	19 Jan 2017	17:30*
	9 Jan 2018	01:45
F6	25 Apr 2011	07:00
	24 May 2011	18:45*
	13 Apr 2012	16:00*
	13 May 2012	02:00*
	11 Jun 2012	15:30*
	11 Jul 2012	02:30
	2 May 2013	13:30*
	1 Jun 2013	01:00*
	19 Jan 2017	19:30*
	9 Jan 2018	03:45

(continued)

(continued)

Drawing ID	Date	Time (UT)
F7	29 Mar 2011	03:45
	27 Apr 2011	16:00*
	17 Mar 2012	11:15*
	16 Apr 2012	01:30*
	15 May 2012	13:30*
	14 Jun 2012	00:45*
	5 Apr 2013	10:15*
	4 May 2013	23:00*
	23 Dec 2016	14:00*
	12 Dec 2017	22:15*
	11 Jan 2018	13:30*

The Mountains of the Moon

"The Moon is not robed in a smooth and polished surface," wrote Galileo, "but is rough and uneven, covered everywhere, just like the Earth's surface, with huge prominences, deep valleys and chasms."

Eager to convey the idea of the Moon as a world in its own right, with a rich and varied landscape full of high mountains and deep valleys, Galileo set about measuring the heights of the lunar mountains. He noticed that in the darkness beyond the terminator (the line of lunar sunrise or sunset, separating the Moon's illuminated half from its dark half), the summits of high mountains caught sunlight when the landscape around them was in shadow. To work out some mountain heights, Galileo observed at the precise time of half Moon (dichotomy) and measured how far an illuminated mountain summit lay from the terminator, choosing an example nearest the center of the Moon to improve accuracy. All that was required was to then apply a little basic geometry in order to calculate the mountain's height.

Galileo used the following example to calculate lunar mountain heights:

Precisely at first quarter phase sunlight fully illuminates the Moon's right-hand side. Suppose an illuminated mountain summit D is visible and measured to be one-twentieth of a lunar diameter on the dark side. In the diagram, the line GCD is a ray from the

Sun, grazing the Moon at the terminator at point C; the base of the mountain, vertically below D, is at A, and E is the Moon's center. The length of DC can be calculated by simply dividing the diameter of the Moon (known since ancient times) by 20. Once this has been calculated, Pythagoras' theorem is applied to the triangle EDC to find ED (the hypotenuse). The height of the mountain, DA, is found by simply subtracting EC (the radius of the Moon) from ED.

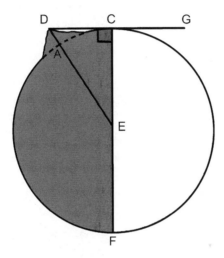

Galileo's method of calculating the height of a lunar mountain (Credit: Peter Grego)

Galileo worked on the assumption that the Moon's diameter was 3,219 km (equal to 2/7 that of Earth), and that the mountain-top he observed at D was a distance of 1/20 the Moon's diameter from C. From this, he calculated that the height of this particular mountain exceeded 6,500 m above the mean lunar surface. Of course, there's no sea level on the Moon to gauge heights by, and it is only very recently that a complete and accurate topographical map of the lunar surface has been completed.

Project #6: Observe and Draw the Moon's Craters

Galileo would be astounded at the advanced level of our knowledge of the Moon. Since the Space Age began, the lunar surface has been mapped by space probes in exquisite detail. Although there remains no significant Moon mapping role left for the amateur observer, those taking the trouble to make lunar drawings will discover an immensely useful and rewarding activity that improves every single aspect of his or her observing skills.

The lunar surface is packed with very fine detail, and the observer's ability to discern this detail improves with time spent at the eyepiece. By concentrating on drawing skills, the observer learns to attend to detail instead of allowing the eye to wander onto the more obvious features. The discipline of accurate lunar drawing will pay off in other fields of amateur astronomy that require pencil work, like planetary observation.

During a course of lunar drawing apprenticeship the apparent confusion of the Moon's landscape, with all its odd tongue-twisting nomenclature, becomes increasingly familiar. It is important to have a degree of confidence in your own drawing abilities; the lunar observer isn't some kind of weird nocturnal art student, and marks aren't given for artistic flair. Observational honesty and accuracy counts above all.

Traditional pencil sketches. A set of soft graphite pencils from HB to 5B and an A5 pad of smooth cartridge paper is essential. If your sketching skills need to be polished, there is no better way than to draw sections of lunar photographs that appear in books and magazines. Try not to be daunted by the sheer wealth of detail visible through the eyepiece. First, it's important to orientate yourself. If you aren't sure what features you are looking at, find your bearings with a good map of the Moon. Your eye will probably be drawn to the terminator, where most relief detail is visible due to the low angle of the sunlight illuminating that region. The area you choose to draw should ideally be quite small, such as an individual crater, and chances are it will be close to the terminator. If a feature you have chosen isn't marked on the map then make a note of nearby features that can be identified and indicate their positions in relation to the unknown area.

Basic outlines are first drawn lightly, using a soft pencil, giving you the chance to erase anything dubious if the need arises. When shading dark areas try to put minimal pressure onto the paper. The darkest areas are ideally shaded in layers, and not in a frenzy of pencil pressure. After several sessions of unpressurized Moon drawing you will surprise yourself at how quickly you see improvement. The most important thing to remember is to be patient. Do not rush, even if you are only practicing.

Set yourself about an hour or two for each drawing session. Patience is vital because a rushed sketch is bound to be inaccurate. Interesting features should be highlighted by short notes. Remember to write down the usual information, such as date, start and finish times (UT), instrument, and seeing.

A sequence showing the development of an observational pencil drawing of the Moon. It depicts the crater La Condamine, as observed on December 2, 1992, between 06:50 and 07:15 UT using a 250 mm Newtonian reflector ×250 (Credit: Peter Grego)

Line drawings. Features can be depicted in simple line form as an alternative to making shaded drawings. Bold lines represent prominent features such as crater rims and the sharp outlines of lunar shadows. The depiction of lunar mountains as upturned "V" shapes should be avoided. If a section of the drawing is full of rough terrain, label it as "rough terrain" – don't attempt to depict by filling the area with a mass of dots and squiggles. Subtle features such as low hills and delicate detail are best recorded with light thin lines. Dashed lines can show features such as rays and dotted lines can mark the boundaries of areas of different tone. Line drawings require plenty of descriptive notes, more so than a tonal drawing, but it has the advantage of being quick and requires the minimum of drawing ability. When done properly, the method can be as accurate and as full of information as any toned pencil drawing.

Cybersketching. The technology now exists to significantly enhance the powers of the visual observer (not to mention increasing the enjoyment factor) by using computers at the eyepiece to produce electronic observational drawings. Cybersketching – using graphics programs and stylus input on the touch screens of PDAs and tablet PCs – is a logical modern offshoot of traditional observing techniques.

Despite their small screen size, PDAs (Personal Digital Assistants) offer much to the visual observer. They're eminently portable, comfortable to hold in the hand for long periods of time, and have a pretty long battery life. Lunar cybersketches made on a PDA may well take up the entire screen area (say 3.5 in. diagonal or smaller), while on the much larger screens of UMPCs (ultramobile PCs) and tablet PCs it's best not to use the whole desktop space when viewed at standard resolution; otherwise a drawing may become unmanageable. Of course, drawings can be scaled to size using these devices, so if starting off a drawing and finding it's either too small to add detail or too large to handle isn't quite the problem that it would be if making a pencil sketch to the wrong scale.

A completed cybersketch of a lunar feature, made on PDA – the crater Abulfeda, observed on 21 September 2008 between 00:10 and 00:55 UT using a 200 mm SCT ×200 (Credit: Peter Grego)

The Sun

It had long been maintained that the Sun's surface was "pure and without blemish." The Church, having wedded itself to certain aspects of ancient Greek philosophy, taught that the heavens were perfect. Yet for many hundreds of years before the invention of the telescope, observers in the far east had been recording spots on the Sun that were often clearly visible with the unaided eye when the Sun was low in the sky or veiled by enough dust, mist, or cloud to enable it to be seen without glare. (Note that this method is not entirely safe; viewing the Sun should never be attempted with the unshielded naked eye.)

Written records of sunspots in Chinese annals date back to 28 BCE; astrologers in ancient China and Korea believed that sunspots, in common with all other heavenly phenomena, were signs of impending happenings on Earth. Western records contain scant references to naked eye sunspots during the same period. In 813 a large naked eye sunspot was widely seen over a period of more than a week; this was considered to be an omen, a superstition reinforced by the subsequent death of Emperor Charlemagne. Another interesting sighting was made by the English monk John of Worcester, who made a drawing of naked-eye sunspots on December 8, 1128; within a week of this, spectacular aurorae were recorded in Korea.

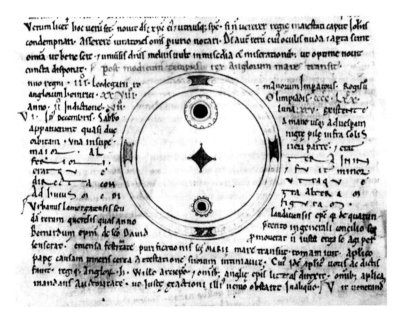

Naked-eye sunspots seen in 1128, recorded in the chronicle of John of Worcester

To digress briefly, the delay between the appearance of a large active sunspot group and auroral activity is now well known; charged particles emitted by active regions on the Sun take several days to reach the vicinity of Earth, where they are guided towards the magnetic poles by Earth's magnetosphere. As these energetic particles collide with atoms in the upper atmosphere, aurorae take place in ring-shaped zones centered over the north and south magnetic poles. However, the link between solar activity and aurorae was not to be explained until the nineteenth century, while direct observation of aurora-producing solar particles first took place in 1954 aboard a high-altitude sounding rocket.

On May 29, 1607, a couple of years before the telescope was first used for astronomy, Johannes Kepler observed a sunspot after having projected the Sun's image through a pinhole onto a paper screen in a darkened room. Kepler mistakenly thought that the spot – which he likened to "a little daub, quite black, like a parched flea" – was a transit of the planet Mercury, since his calculations had suggested that the innermost planet would lie directly between Earth and the Sun on that day. Kepler published his observation as such in his *Treatise on Comets* (1607) and at length in *Extraordinary Phenomenon, Mercury Seen Against the Sun* (1609), but he was eventually to realize his mistake.

Thomas Harriot made the first telescopic observations of sunspots by wearing dark colored glasses in an attempt to cut down on the glare of the image; not only was the quality of the view diminished, the technique was extremely unsafe and risked permanently blinding the observer. Later, Harriot chose to view the Sun when it was naturally filtered by cloud or mist – still a very unsafe technique, the risks of which were not fully appreciated by the scientist. Between December 8, 1610, and January 18, 1613, he succeeded in making 199 sunspot observations. Sometimes Harriot was unwise enough to view the Sun without even these most basic precautions; one of his solar observations, made on February 17, 1612, indicates that he suffered some form of retinal damage: "All the sky being cleare, and the sonne, I saw the great spot with 10 and 20, but no more, my sight was after dim for an houre."

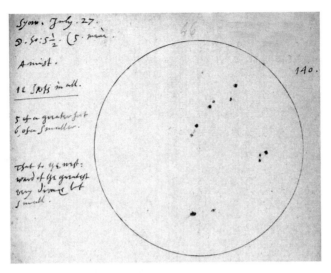

One of Thomas Harriot's telescopic solar observations of 1609

After developing the far safer technique of camera-obscura telescopy to observe the Sun's disk by means of projection, the German father-son team of David Fabricius (1564–1617) and Johannes Fabricius (1587–1616) noted the movements of sunspots across the Sun's disk; spots would always appear at the eastern edge of the Sun, traversing the disk to disappear at the western edge in around a fortnight. After an extensive series of observations they concluded that sunspots were physically attached to the Sun's surface, and deduced the Sun's period of axial rotation. In 1611 Johannes published his observations and theories in *De Maculis in Sole Observatis, et Apparente earum cum Sole Conversione Narratio* (*Narration on Spots Observed on the Sun and Their Apparent Rotation with the Sun*), but the book was not widely read at the time.

Working independently, without a knowledge of Fabricius' research, the Jesuit priest Christopher Scheiner (1573–1650) first observed sunspots in 1611 while in post as a lecturer on astronomy at Ingoldstadt. Under the impression that he was first to have discovered sunspots, Scheiner published his observations a year later in *Letters* under the pseudonym of Apelles. At this time, Scheiner mistakenly thought that sunspots were tiny planets orbiting near the Sun and seen as silhouettes in transit. He went on to make more than 2,000 solar observations over the next four decades, during which time he refined his methods and developed a simple form of equatorial mount that helped in his observing.

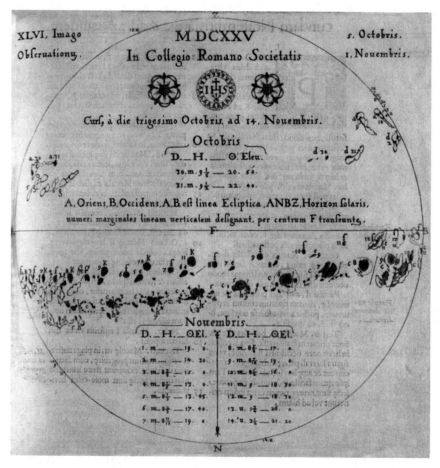

Scheiner's illustration of the Sun's disk and the movement and development of sunspots between October 30 and November 14, 1630

It's not known when Galileo's observations of the Sun began, but he was not at all happy that Scheiner claimed to be the first to have observed sunspots. Galileo maintained that he had viewed these solar features long before Scheiner, and that it was his opinion that sunspots were attached to the Sun, contrary to Scheiner's belief expressed in *Letters*. There was to follow an acrimonious debate over who was to be credited with the discovery of sunspots. Scheiner's monumental work on the Sun, titled *Rosa Ursina sive*

Sol (*The Sun, Rose of Orsini,* 1630) maintains that he was the discoverer of sunspots, but does acknowledge Galileo's original contention that they are part of the Sun's surface; *Rosa Ursina* is remarkable in that it contains a great many personal attacks upon Galileo. Scheiner was by no means the first or last to take issue with the great man.

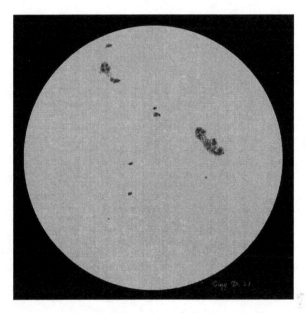

One of Galileo's observational drawings of the Sun, made in 1612

Project #7: Observe the Sun

If you take the right precautions, you can observe sunspots for yourself using a small, steadily held telescope.

Safety first! It's important NEVER to look directly at the Sun through a telescope or binoculars (or even except very briefly with the naked eye). The Sun is a very dangerous object to view at any time. Your eyesight is at risk – possibly permanent blindness – if you attempt to observe the Sun without taking the proper precautions.

The best and safest way to observe the Sun is to project its image onto a shielded white screen. If your telescope has a finder (a small secondary telescope), make sure to keep its lenses covered and never attempt to locate the Sun using it. Telescopes trained on the Sun should never be left unattended; the Sun's focused image is very hot and can easily melt components, or the telescope may be inadvertently looked through by inexperienced people. Small dark filters that fit onto eyepieces must never be used. These can crack without warning and damage your eyes. Never use photographic film, smoked glass, or a shiny CD as a filter.

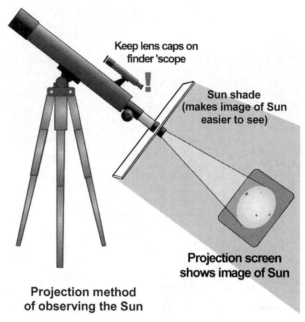

Projection method of observing the Sun

The safest method of observing the Sun is to project its image through a small telescope onto a shielded white card (Credit: Peter Grego)

The number of visible sunspots varies from day to day, the average numbers increasing and decreasing over a cycle. From one maximum to another, the solar cycle is roughly 11 years long. The cycle has been observed, on and off, since 1610, and cycles have been designated numbers since 1755. The last solar maximum took place in 2001, but the solar cycle that followed – cycle number 23 – began in April 1996 and lasted more than 13 years, making it the longest since cycle number 4 (1784–May 1798). The year 2013/2014 may see the next solar maximum, with a minimum perhaps around 2018/2019.

The Sun rotates on its axis in the same direction as the planets orbit the Sun – anticlockwise, as viewed from above the Sun's north pole. A ball of gas, the Sun's rotational period is differential, varying between 25.6 days at the equator to 30.9 days at a latitude of 60° N or S (and even slower towards the poles). Sunspots first appear at the Sun's eastern edge, transit the solar meridian more than 6 days later, and disappear beyond the western limb after another 6 days or so.

To observe sunspots, your telescope needs to be held on a sturdy mount; the shielding card goes around the base of the telescope tube, while the projection card needs to be held quite firmly, too. Projection screens to fit onto standard makes of telescopes can be bought from most astronomical retailers; alternatively, one can be constructed fairly easily using a lightweight frame. The telescope is pointed in the Sun's direction by aligning the shadow of the tube on the white shielding card. It's best to use a budget eyepiece (one that you are prepared to risk being exposed to the Sun's focused heat and light) with a low magnification in order to be able to project the entire disk of the Sun onto the card. Once found, the image is focused by making the edge of the Sun appear as sharp as possible.

What can be seen in a white light projected image of the Sun? Sunspots are the most obvious signs of solar activity visible in white light. These features usually develop from small dark pores in the photosphere and can take several days to grow to maximum size – on average, around 10,000 km in diameter. Fully formed sunspots often have two components – a darker central area called the umbra and a dusky peripheral region called the penumbra. Sunspot behavior can be unpredictable; some decay after a day or two on the solar disk, while others can grow to enormous proportions – as large as 150,000 km across. Large sunspots may last for many weeks, long enough for one or more rotations of the Sun.

Sunspots are caused by disturbances within the powerful magnetic field generated within the Sun. The difference in temperature and brightness between the 5,500°C photosphere (the white visible surface) and a sunspot's 3,500°C interior makes the sunspot appear dark. If a sunspot were somehow removed from the solar surface, it would appear as a dazzlingly brilliant object.

In white light the brightness of the Sun appears to diminish towards the edge of the solar disk. This phenomenon, called limb darkening, is produced because the Sun's temperature and the amount of light it emits increases with depth. On looking towards the center of the Sun's disk, we're viewing the deepest and hottest layers within the Sun – those that emit the most light; towards the Sun's edge, only the upper, cooler, dimmer layers are visible. The presence of limb darkening is proof that the Sun is not a solid sphere of burning material.

If you're fortunate you may even see faculae (meaning 'little torches'), which are bright areas near the Sun's edge, often running in lengthy sinuous bands; limb darkening makes them easier to see in white light. The origin of faculae is not well-understood, but they are undoubtedly produced by magnetic phenomena.

Photography of sunspots on the projected solar disk over the course of a few days will show you how speedily the Sun rotates upon its axis and how the spots develop over time. When imaging the Sun, attempt to take your picture as close to the optical axis of the instrument as possible, as square-on to the projection screen as you can manage; otherwise the Sun may appear distinctly egg-shaped. Use optical or digital zoom with your camera to zero-in on the sunspots for a more detailed view of them. Image enhancement on the computer will bring out the limb darkening nicely, along with any faculae that may be present.

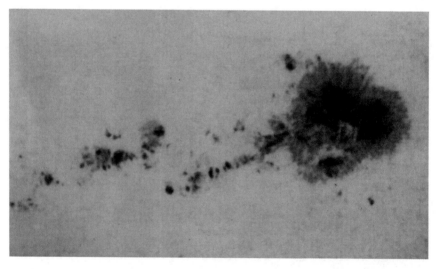

Sunspot group photographed on October 25, 1969 (Credit: Ramon Lane)

Planets

Galileo's observations of the planets were startling. Instead of appearing as bright points of light like the stars, the telescope revealed them to be tiny disks; the planet Venus even showed a series of phases, just like the Moon does. This couldn't be explained using the Ptolemaic system. Furthermore, according to the ideas of Ptolemy and the doctrine of the Church, Earth was at the center of the Universe and all bodies orbited Earth. However, between January 7 and 13, 1610, Galileo observed four satellites orbiting the planet Jupiter. Here was more clear and irrefutable proof that the geocentric theory was wrong.

Venus's Phases

With the exception of transient phenomena such as a really brilliant comet or a swiftly moving fireball, Venus is the third brightest celestial object. Second only in brightness to the Sun and the Moon, the planet's dazzling presence in the evening or morning skies often attracts the attention of non-astronomers.

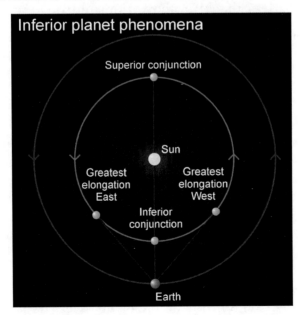

Diagram showing the orbit and phenomena displayed by Mercury or Venus (not to scale) (Credit: Peter Grego)

A so-called inferior planet, Venus orbits the Sun nearer than Earth does, and as a consequence it displays a complete sequence of phases. When Venus is at its furthest from Earth, on the other side of the Sun, it is at superior conjunction and unobservable. Within a month or so of superior conjunction, Venus has edged east of the Sun far enough to be visible with the unaided eye in the evening skies. It takes Venus about 7 months to go from superior conjunction to reach maximum eastern elongation, between 45° and 47° from the Sun – far enough for the planet to be seen against a truly dark evening sky, several hours after sunset.

Some apparitions of Venus are more favorable than others, depending on how high the planet is above the horizon at sunset. Maximum eastern elongations are most favorable from northern temperate latitudes during the spring, when the planet can be as high as 40° above the western horizon at sunset and observable for more than four hours in darkening evening skies. For observers in northern temperate regions, unfavorable eastern elongations take place during the autumn, when the angle made by Venus, the set-ting Sun, and the horizon is at its smallest. On these occasions, Venus can be less than 10° above the southwestern horizon at sunset, and consequently the telescopic image of Venus is liable to suffer greatly from the effects of atmospheric turbulence.

Venus's journey back towards the Sun is accomplished much faster than its outward leg, taking just 10 weeks or so to move from maximum eastern elongation to inferior conjunction, when it is nearest Earth in space. Venus' maximum brightness in the evening skies, when it gleams a brilliant magnitude –4.3, is reached around 36 days before inferior conjunction, at an elongation of around 39° east of the Sun; it is then a 25% illuminated crescent with an apparent angular diameter of around 40 arcseconds. This is just about large enough for the crescent to be resolved by those with excellent eyesight; there are numerous reports of naked-eye sight-ings of the crescent Venus, but none that are particularly convinc-ing from the pre-telescopic era.

Following inferior conjunction, Venus makes its presence known west of the Sun in the pre-dawn skies. As with evening apparitions, maximum brightness takes place around 36 days into the apparition, when Venus is around 39° east of the Sun, a crescent phase around a quarter illuminated and shining at magnitude –4.3. It takes Venus around 10 weeks to reach maximum elongation west of the Sun. For observers in northern temperate latitudes, maximum western elongations are at their most favorable when they occur during the autumn, when Venus rises above the eastern horizon almost 5 h before the Sun. Maximum western elongations that take place during the spring are poor, with Venus barely visible above the southeastern horizon before dawn. Venus's journey back towards the Sun and its return to superior conjunction takes around 7 months.

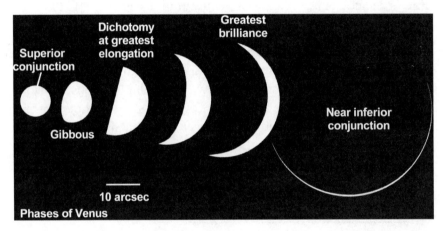

The phases of Venus can be seen through a small telescope (Credit: Peter Grego)

Galileo made the first known observations of the phases of Venus towards the end of 1610. He clearly observed the planet's phases and change in apparent angular diameter, phenomena that were prohibited by the Ptolemaic system (which would never allow Venus to be anything more than half-illuminated from the perspective of Earth). These observations essentially ruled out the Ptolemaic system and was compatible only with the Copernican system and the Tychonic system (and other assorted geoheliocentric models such as the Capellan and Riccioli's extended Capellan model).

Galileo's depiction of the phases of Venus, from *Il saggiatore* (1623)

Looking at the phases of Venus from the viewpoints of both the Copernican (heliocentric) and Ptolemaic (geocentric) systems, it can be seen that only the Copernican view can explain the observations. The Ptolemaic system would allow Venus to show either as a thin crescent if the planet orbited Earth closer than the Sun or, in the case that it orbited further away than the Sun, it would only show a full or gibbous disc. Contrast that with the Copernican view where Venus orbits the Sun and shows a complete range of phases and a variation in its apparent diameter depending on where it is in its orbit.

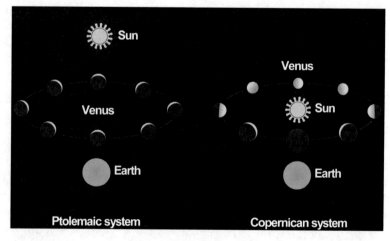

Phases of Venus in the Ptolemaic and Copernican systems (Credit: Peter Grego)

Project #8: Observe Venus's Phases

Data for Venus 2011–2020

Greatest western elongation (morning)		Greatest eastern elongation (evening)	
8 January 2011	47.5°	27 March 2012	46.5°
15 August 2012	45.5°	1 November 2013	47.5°
22 March 2014	46.5°	6 June 2015	45.5°
26 October 2015	46.5°	12 January 2017	47.5°
3 June 2017	45.5°	17 August 2018	45.5°
6 January 2019	47.0°	24 March 2020	46.0°

Jupiter

On January 7, 1610, Galileo was astonished to observe several tiny points of light in the immediate vicinity of Jupiter; his amazement grew during the ensuing weeks when he saw that these four luminous specks were moons orbiting Jupiter, just like a Solar System in miniature. The four satellites – Io, Europa, Ganymede, and Callisto, in order from Jupiter – are often called the Galilean moons in honor of their discoverer.

At first, Galileo named his discovery the *Cosmica sidera* (the Cosmian stars), after his patron Cosimo de' Medici, Grand Duke of Tuscany, but following advice from the Grand Duke's secretary chose to call them the *Medicea sidera* (the Medicean stars), honoring the four powerful Medici brothers. We have Simon Marius (1573–1624), a contemporary of Galileo's and rival claimant to the discovery of Jupiter's moons, to thank for the classical names that we use today, which first appeared in his *Mundus Jovialis* (1614). The discovery was firm proof that not all celestial bodies orbited Earth.

Project #9: Observe the Galilean Moons

Satellite	Dia (km)	Orbit (km)	Period (days)	Max apparent magnitude
Io	3,642	422,000	1.769	5.0
Europa	3,122	671,000	3.551	5.3
Ganymede	5,262	1,070,000	7.154	4.6
Callisto	4,820	1,883,000	16.689	5.7

The Galilean moons can be recorded night after night using a small telescope or a good pair of sturdily mounted binoculars. The moons appear to shuffle back and forth along a line parallel with Jupiter's equator, which is always roughly face-on to us. By estimating the relative brightness of each satellite it's possible to identify them without consulting an astronomical ephemeris, simply by virtue of their brightness. As you can see from the table, Ganymede is the brightest satellite.

Like Galileo, you can estimate the distance each satellite is from Jupiter in terms of the planet's apparent diameter. As the equatorial diameter of Jupiter is about 143,000 km the satellite's maximum distance from Jupiter in terms of Jupiter's apparent diameter will be Io (2.5), Europa (4.2), Ganymede (7) and Callisto (13). Through a larger telescope it's possible to observe eclipses of the moons as they plunge into Jupiter's shadow or behind its bright limb, transits of the moons and their shadows across the Jovian disk, and under certain circumstances the moons will mutually eclipse and occult one another.

A high magnification view of the Galilean moons, as seen through a 150 mm refractor (Credit: Peter Grego)

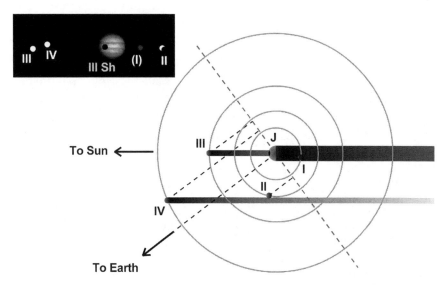

Various transit, eclipse, and occultation phenomena of the Galilean moons (Credit: Peter Grego)

Dynamic Gas Giant

As if the Galilean moons weren't fascinating enough, the planet itself is a joy to observe through a reasonably sized telescope. Jupiter, a gas giant without a solid surface, is noticeably oblate, its broadness at the equator caused by its rapid rate of spin; each revolution of the largest planet in the Solar System takes a little less than 10 h.

Jupiter's upper atmosphere is in constant turmoil, and no truly permanent features exist on the planet. Banding of the planet's atmospheric belts and zones, produced by the planet's rapid spin, can easily be seen through the telescope. The dark belts and light zones vary in intensity from year to year, but the most prominent are usually the North and South Equatorial Belts. Features within the clouds change from week to week, as spots, ovals, and festoons develop, drift in longitude, interact with one another, and fade away. Features usually remain within their own belt or zone, never changing latitude.

Perhaps the nearest thing that Jupiter has to a "permanent" feature is the famous Great Red Spot, a giant anticyclone that could easily swallow Earth. Having wandered around the planet's South Tropical Zone since at least the mid-nineteenth century, the Great Red Spot varies in intensity from year to year, ranging from a barely discernable gray smudge to a sharply defined brick red oval easily visible through small telescopes.

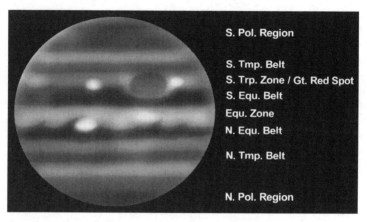

Jupiter's main belts and zones (Credit: Peter Grego)

| | Oppositions of Jupiter | |
Date	Constellation	Declination
29 October 2011	Aries	+11° 53′
3 December 2012	Taurus	+21° 20′
5 January 2014	Gemini	+22° 40′
7 February 2015	Cancer	+16° 27′
8 March 2016	Leo	+05° 58′
8 April 2017	Virgo	−05° 41′
9 May 2018	Libra	−16° 04′
10 June 2019	Ophiuchus	−22° 26′
14 July 2020	Sagittarius	−21° 54′
20 August 2021	Capricornus	−13° 32′

An Amazing Conjunction

Conjunctions occur when two or more astronomical bodies appear to move close to each other in the night sky. The Moon often finds itself in conjunction with the planets, and the planets themselves undergo conjunctions with one (or, more rarely, two) other planets. Conjunctions are just line-of-sight phenomena, but they often make superb visual or telescopic spectacles, presenting great opportunities to snap brilliant astronomical photographs.

As a watcher of the night skies, Galileo was familiar with lunar and mutual planetary conjunctions. However, one conjunction in particular – a remarkably close conjunction between Jupiter and Neptune – went by completely unnoticed by the great scientist, despite the fact that he observed the event telescopically and recorded it in his notes. Of course, both outer planets Uranus and Neptune were completely unknown and unsuspected in Galileo's day; Uranus was discovered by William Herschel (1738–1822) on March 13, 1781, while Neptune was found by Johann Galle (1812–1910) on September 23, 1846, as a result of calculations by Urbain Le Verrier (1811–1877).

Galileo observed Jupiter and Neptune in the same telescopic field of view on three occasions, first on December 28, 1612, again on the following evening, and finally on January 27, 1613. Neptune shines at magnitude 7.9 and is too faint to be seen with the naked eye; it presents a disk averaging just 2.3 arcseconds across. On the basis of a single observation Galileo could not have hoped to realize that the faint "star" near Jupiter was a sister planet. However, Galileo was a careful and accurate observer; his plots of the positions of the Jovian satellites were literally "spot on." On his third observation with Neptune in the same field as Jupiter, he commented that the faint "star" appeared to have been in a slightly different position before. It's unfortunate that he didn't follow up on this suspicion, but Neptune was beginning to near the edge of the Jovian field of view by the time of his final observation; had it been nearer Jupiter, the 'star' may have been identified as displaying a movement against the celestial sphere, something that no star appears to do in such short a period of time.

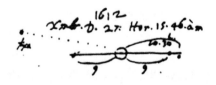

Galileo unwittingly observed Neptune on three occasions, thinking it was a "fixed star." This is his first observation, made on December 27, 1612, compared with a computer simulation of the same scene (Credit: Peter Grego)

On January 3–4, 1613, Jupiter went on to occult Neptune altogether – an exceedingly rare phenomenon. Had Galileo observed this event, he would have realized that the 'star' being hidden was not merely a pinpoint of light; its protracted fading or brightening (disappearance or reappearance) at the edge of Jupiter of more than half an hour meant that it had an appreciable diameter. Knowing how fast Jupiter moved, Galileo would have been able to estimate the apparent diameter of Neptune. A knowledge of the angular velocity of Neptune would also have enabled its distance from the Sun to be calculated according to Kepler's laws.

Project #10: Observing Opportunity: Significant Conjunctions 2011–2021

Close planetary approaches and conjunctions are fascinating to view with the unaided eye and through binoculars. When observed through a telescope, the appearance of two or more distinct objects in the same field of view seems to add depth to the Solar System. The actual field of view presented by any combination of telescope and eyepiece depends on the focal length of the telescope, the focal length of the eyepiece, and the eyepiece's apparent field of view. Short focal length telescopes in combination with eyepieces of long focal length and wide fields deliver views of the broadest areas of sky.

Date	Objects	Separation
19 April 2011	Mercury and Mars	0.8°
1 May 2011	Mars and Jupiter	0.4°
11 May 2011	Venus and Jupiter	0.7°
16 May 2011	Mercury, Venus, Mars, and Jupiter	Within 5°
23 May 2011	Venus and Mars	1°
30 June 2011	Moon and Venus	0.7°
22 May 2012	Mercury and Jupiter	0.4°
17 June 2012	Moon and Jupiter	0.5°
27 November 2012	Venus and Saturn	0.6°
8 February 2013	Mercury and Mars	0.4°
7 April 2013	Venus and Mars	0.6°
8 May 2013	Mercury and Mars	0.5°
26 May 2013	Mercury, Venus, and Jupiter	Less than 3°
22 July 2013	Mars and Jupiter	0.8°
3 November 2013	Moon and Mercury	0.5°
26 November 2013	Mercury and Saturn	0.4°
28 April 2014	Moon and Saturn	0.9°
18 August 2014	Venus and Jupiter	0.2°
11 January 2015	Mercury and Venus	0.6°
22 February 2015	Venus and Mars	0.4°
1 July 2015	Venus and Jupiter	0.3°
16 July 2015	Mercury and Mars	0.2°
7 August 2015	Mercury and Jupiter	0.5°
18 October 2015	Mars and Jupiter	0.4°
3 November 2015	Venus and Mars	0.7°
9 January 2016	Venus and Saturn	5 arc min
13 May 2016	Mercury and Venus	26 arc min
27 August 2016	Venus and Jupiter	4 arc min
28 June 2017	Mercury and Mars	47 arc min
16 September 2017	Mercury and Mars	4 arc min
28 March 2020	Mars Jupiter and Saturn	4°
21 December 2020	Jupiter and Saturn	6 arc min
10 January 2021	Mercury, Jupiter, and Saturn	All within 2°

Saturn

Galileo saw that the five known planets all showed disks – they were appreciable worlds in their own right, not simply moving stars – but one planet caused him certain problems of interpretation. Through his telescope, Saturn displayed odd little protrusions on each side of its main disk, an appearance which made him suspect that the planet consisted of one large central planet with two smaller ones on each side. He wrote: "I have observed the highest planet [Saturn] to be tripled-bodied. This is to say that to my very great amazement Saturn was seen to me to be not a single star, but three together, which almost touch each other."

Yet when he came to observe Saturn again 2 years later in 1612, Saturn's appearance had changed. To his astonishment, the little blobs had disappeared, leaving just a single disk on view. He wrote: "I do not know what to say in a case so surprising, so unlooked for and so novel."

By 1616 the small protrusions had reappeared, leaving Galileo totally perplexed. He eventually settled on an explanation that he was not entirely satisfied with – that Saturn was bestowed with peculiar "arms" or "ears" that somehow grew and disappeared over time. "The two companions are no longer two small perfectly round globes," he wrote, "but are at present much larger and no longer round ... that is, two half ellipses with two little dark triangles in the middle of the figure and contiguous to the middle globe of Saturn, which is seen, as always, perfectly round."

Galileo's observations of Saturn in 1610 (*top*) and 1616

Galileo's telescope was not powerful enough for him to discern the true cause of mysterious morphing Saturn. Half a century was to elapse before the right explanation was given by the Dutch astronomer Christiaan Huygens, whose more powerful telescope enabled him to discern that Saturn was surrounded by a thin, flat ring system, "nowhere touching the globe and inclined to the ecliptic."

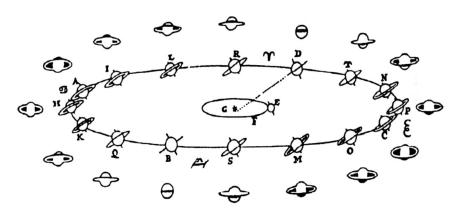

Huygens' explanation of the changing appearance of Saturn – a ring system whose tilt towards the Earth varies during the course of its orbit around the Sun

Project #11: Following Saturn

Saturn takes 29 years to complete each orbit around the Sun. For about 14 years the north face of the planet and rings is presented towards us, followed by a short period when the rings are edge on, followed by 14 years when the planet's southern aspect is on view. From 2010 onwards, the tilt of Saturn's rings towards Earth gradually increases as the planet's north pole slowly inclines towards us; their tilt slowly increases to a maximum in October 2017. However, offsetting the increasingly pleasant aspect of Saturn's rings is the planet's increasingly low declination. It hits a deep southern low declination in Sagittarius in 2018, which means that observers in northern temperate regions won't get such a good view.

| Date | Oppositions of Saturn | | |
	Ring tilt	Constellation	Declination
4 April 2011	8.6°	Virgo	−02° 57'
15 April 2012	13.3°	Virgo	−07° 31'
28 April 2013	17.5°	Virgo	−11° 42'
10 May 2014	21.7°	Libra	−15° 20'
23 May 2015	24.3°	Libra	−18° 20'
3 June 2016	26.0°	Ophiuchus	−20° 34'
15 June 2017	26.6°	Ophiuchus	−21° 58'
27 June 2018	26.0°	Sagittarius	−22° 27'
9 July 2019	24.4°	Sagittarius	−22° 00'
20 July 2020	21.6°	Sagittarius	−20° 39'
2 August 2021	18.1°	Capricornus	−18° 26'

Stars

Galileo published many of his startling observations in early March 1610 in *Sidereus Nuncius* (*The Starry Messenger*), which included his observations of the Milky Way, stars in Orion, and the open clusters of Praesepe and the Pleiades. Galileo found that there were many hundreds of stars clearly visible in his telescope that were quite invisible to the naked eye. His 30 mm telescope would have shown him around eight times the number of stars visible with the naked eye.

Under very good, dark, moonless conditions, it's possible to see around 4,000 stars in the northern and southern hemispheres combined. Below is a table showing approximately how many stars can be observed in the whole sky, depending on the size of your telescope. The limit for ordinary 10×50 binoculars is about magnitude 9.5, allowing some 120,000 stars to be seen.

Size of objective	Limiting magnitude	Number of stars visible
Naked eye ~ 8 mm	6.0	4,000
50 mm	10.4	500,000
75 mm	11.3	1,400,000
100 mm	11.9	2,700,000
125 mm	12.4	5,400,000
150 mm	12.8	8,400,000
175 mm	13.1	11,600,000
200 mm	13.4	16,100,000

Thus, a 200 mm telescope will show about 4,000 times more stars than the naked eye.

Galileo observed that the bright open star cluster of the Pleiades (known as the Seven Sisters) in Taurus consisted of dozens of stars that can't be seen without optical aid; his drawing of 1610 depicts stars. Praesepe (known as the Beehive), a large open cluster in Cancer, appears as a nebulous patch with the unaided eye, but Galileo resolved it into dozens of individual stars. Similarly, he saw that the misty band of the Milky Way was formed out of thousands of stars that were too dim to be seen individually. When we view the Milky Way we are looking directly along the plane of our own Galaxy; perspective produces the impression that the stars are crowded together, but there's an average of a good few light years between each star in our Galactic neck of the woods.

Galileo's observational drawing of the Pleiades, compared with a telescopic image. Galileo marked the brighter, previously known stars with an outline shape

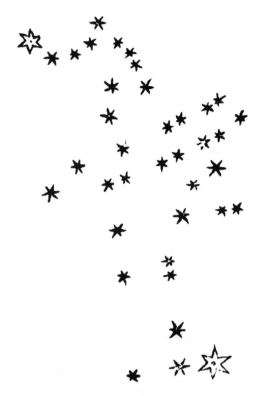

Galileo's observation of Praesepe, an open cluster in Cancer

Project #12: Pleiades Perception

Try to draw what you can see of the Pleiades using the unaided eye. A person with good eyesight on a moonless night might glimpse six, seven, or eight stars; under excellent conditions; those with exceptional vision may glimpse a dozen or so. How many stars in the Pleiades can you see using binoculars or through the eyepiece of a small telescope at a low magnification? How is it best to go about charting the stars you can see, as accurately as possible, using just your eye and a sketchpad?

The Final Years

In terms of observing and recording astronomical phenomena, the period between 1609 and 1612 was remarkably fruitful for Galileo. But just as he was becoming one of the western world's most famous scientists the storm clouds were forming around him.

Galileo's famous struggle with the Church really began around 1612. A series of events instituted by the Church, known as the Inquisition, was under way, and its purpose was to ensure that the Church's doctrines were being adhered to; an exceedingly dim view was taken of all forms of blasphemy and heresy. Galileo's battle culminated with a trial in April 1633 in which he was forced to abjure, curse, and detest the heresy that he supported and taught, including the Copernican view that Earth moved around a motionless Sun. Placed under house arrest and forced to recite penitential psalms every day for 3 years, he was not even allowed out to walk in his garden.

During his final years Galileo worked on his greatest book, *Discourses on Two New Sciences*. In this book, published in the Netherlands in 1638, he set out all the observations and theories he had worked on over the previous 40 years. He developed kinematics and mechanics, describing the motion of bodies free from frictional forces. He was able to extrapolate from experiments that without frictional forces a body would keep on moving forever.

By 1637 Galileo's eyesight was failing. Those direct visual observations of the Sun that he had made decades earlier certainly didn't help his visual acuity. First he lost his sight in his right eye, which was so important in his observations; by December 1637, after losing sight in his left eye, he was left completely blind. In his later years Galileo was helped by Vincenzo Viviani, a 16-year-old acolyte who came to live with him in his home in Arcetri and who acted as his scribe and wrote his first biography.

However, Galileo hadn't quite finished his scientific career. In 1637 he observed and wrote about his discovery of lunar libration, a slow, apparent oscillation of the Moon as it orbits Earth, enabling us to peer a little beyond the mean edges of the lunar disk. The phenomenon is mainly due to the Moon being in a synchronous orbit – taking exactly one revolution around Earth to make one turn on its own axis. But being in a slightly elliptical orbit, each month the Moon's Earth-turned aspect appears to slowly oscillate.

In 1638 Galileo devised a method to measure the speed of light. His idea was to place two observers a few kilometers apart, each holding a lantern at night. They would start with both lanterns covered; one observer would open his shutter, releasing a light beam, and when this was seen by the second observer he would open the shutter of his lantern. Knowing the time difference and distance of separation of the observers would enable the speed

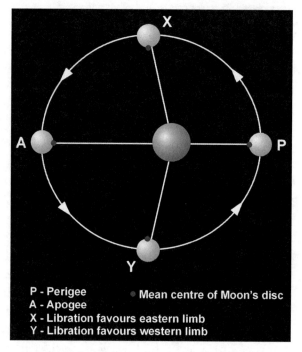

P - Perigee   • Mean centre of Moon's disc
A - Apogee
X - Libration favours eastern limb
Y - Libration favours western limb

This shows how libration affects the apparent position of features on the Moon's near side. A total of 59% of the Moon may be viewed over time through libration (Credit: Peter Grego)

of light to be calculated. Of course, at 300,000 km/s the time required by light to travel to and from observers separated by 3 km would be 20 millionths of a second – slightly less than two peoples' reaction times of 0.4 s! So, this idea would never have produced a reliable result. But just 33 years after Galileo's death, the Danish astronomer Ole Roemer used timings of the Galilean moons to obtain an accurate value of the speed of light. Using the values of the Earth-Sun distance at the time, Roemer's observations in 1675 showed the speed of light was approximately 220,000 km/s.

Galileo was a genius who transformed the way we look at the world and how we can discover the world through science. He insisted on performing experiments and making careful observations, using mathematics as the language of science. He showed that the careful quantitative measurement of motion could enable humankind to deduce the laws of nature. Galileo developed the telescope and microscope, experimented with pendulums and sound, wrote about the tides, developed numerous instruments such as the proportional compass and hydrostatic balance, and investigated the strength of materials. Moreover, he was highly effective at publishing his ideas throughout Europe and was tremendously brave in the face of ignorance, opposition, and criticism, especially that which was leveled at him by the Church. He was a great observer, a highly skilled instrument maker, and an ingenious inventor. Galileo saw what nobody had seen before, and in doing so overturned the traditional ideas of Aristotle and fatally wounded established dogma. The way was paved for the next giant of science, Isaac Newton.

3. Newton's Universe

In 1666 the sleepy little hamlet near Grantham in Lincolnshire, eastern England, hosted a world-changing event centering around one remarkable person. History tells us that this was by no means an "ordinary" person – although in many respects he was just as fallible and human as any other mortal. It's a measure of the indelible impact that this person had on the world that his name is not only remembered in history lessons but lives on in the fields of mathematics, physics, astronomy, and optics to a greater extent even than that of Galileo. That extraordinary human being was Isaac Newton.

Born into a farming family at Woolsthorpe Manor on Christmas Day 1642 – less than a year after the death of Galileo – Isaac Newton was named after his recently deceased father. Many weeks premature, the baby was fortunate to have survived in an age where infant mortality rates were extremely high. These were difficult times for the family. His mother Hannah and grandmother Margery struggled to care for the child in addition to running the farm, all against the tumultuous backdrop of the English Civil War (1641–1651).

When Newton was three, his mother married a prosperous minister from a nearby village; the boy remained at Woolsthorpe with his grandmother but stayed in contact with his mother. Fortunately his stepfather's library was ample, and Newton would spend much time absorbed in reading, devouring knowledge as fast as his eyes could scan the printed page.

Growing up an intensely curious child, young Newton's desire to understand the world around him was noticed by his uncle, William Ayscough, who advised that he be sent to the nearby King's School in Grantham rather than tend to duties on the family farm. Yet on her return to Woolsthorpe in 1656, following the death of her second husband, Hannah withdrew her son from school in the hope of transforming him into a farmer. Fortunately the boy showed little inclination for the farming life, and his mother was eventually

P. Grego and D. Mannion, *Galileo and 400 Years of Telescopic Astronomy*,
Astronomers' Universe, DOI 10.1007/978-1-4419-5592-0_3,

persuaded by Newton's former teacher at Grantham to allow him to prepare for his entrance to Cambridge University.

It was a wise decision, for in just a few short years Newton was able to establish himself as a brilliant physicist and mathematician whose remarkably perspicacious insights made the Universe understandable through a set of universal laws. His optical discoveries were ultimately to lead to the creation of the world's greatest telescopes, while his research laid the foundations of modern physics, giving scientists powerful means by which the Universe could be seen, investigated, and understood.

Sir Isaac Newton in 1689, painted by Godfrey Kneller

Exciting Times

Isaac Newton's formative years were exciting ones in the history of science. In 1660, just a year before he had begun his degree at Trinity College, Cambridge, an organization dedicated to the new idea of "experimental philosophy" was founded. Such an organization, whose purpose was to encourage scientific research and disseminate new observations, discoveries, and theories, had originally been suggested by Sir Francis Bacon (1561–1626), but only now, with the convergence of great intellects whose common desire was to belong to a thriving scientific community at the heart of Restoration England, did the idea take root. Founding members of this organization – sometimes referred to as the "Invisible College" – included such notables as Christopher Wren (who had lectured on astronomy on its inaugural evening), Robert Boyle, and Robert Hooke.

Recognizing the enormous merits of the new group, King Charles II bestowed his royal seal of approval on the organization in 1662 so that it became "The Royal Society of London for the Improvement of Natural Knowledge." Now known simply as the Royal Society, the organization has been going strong for three and a half centuries. Newton himself became a member of the Royal Society in 1672 and served as its president between 1703 and his death in 1727.

It was against this background of a burgeoning new approach to science that Newton began to display his brilliant insights into nature. The young Cambridge scholar became fascinated by optics, light, and color after reading accounts of investigations by the likes of Boyle and Hooke. Published under the aegis of the Royal Society, Hooke's famous book *Micrographia* (1664) details observations made through various lenses, ranging from (mainly) microscopical to astronomical objects. In it, Hooke writes:

> By the means of Telescopes, there is nothing so far distant but may be represented to our view; and by the help of Microscopes, there is nothing so small, as to escape our inquiry; hence there is a new visible World discovered to the understanding. By this means the Heavens are open'd, and a vast number of new Stars, and new Motions, and new Productions appear in them, to which all the ancient Astronomers were utterly Strangers. By this the Earth it self, which lyes so neer us, under our feet, shews quite a new thing to us, and in every little particle of its matter; we now behold almost as great a variety of Creatures, as we were able before to reckon up in the whole Universe it self.

Newton was perhaps also spurred on in his optical research by Hooke's claim in *Micrographia* that "As for Telescopes, the only improvement they seem capable of, is the increasing of their

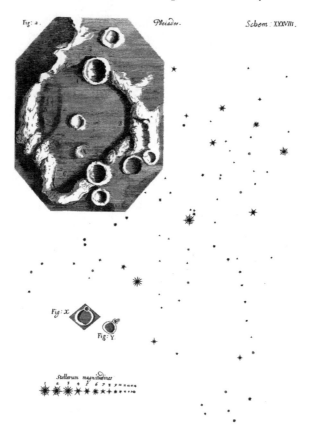

Hooke's observations of the Pleiades and the lunar crater Hipparchus, published in *Micrographia*

length" – a claim that was to be overturned in a most dramatic fashion by Newton, as we shall see.

Newton's Annus Mirabilis

In late 1665, soon after Newton had gained a somewhat mediocre bachelor's degree from Trinity College, Cambridge, the university was forced to shut its doors for the next 2 years because of the Great Plague. Newton returned home to the relative rural safety of Woolsthorpe, where, during the course of the next 18 months, he experienced a most extraordinary series of scientific revelations. Of this remarkable period, Newton wrote: "...in those days I was in my prime of age for invention, and minded mathematics and philosophy more than at any time since."

The year 1666 – forever infamous for being the year in which the Great Fire destroyed most of London – proved to be Newton's *Annus Mirabilis*. While at Woolsthorpe he conceived and developed the binomial theorem and calculus in mathematics, began to formulate his laws of motion and his law of universal gravitation in physics, while in optics he developed our understanding of light and invented the reflecting telescope – the basis of all the world's largest telescopes. Newtonian mechanics dominated physics for more than 200 years, and although Einstein's general theory of relativity (1915) has transcended Newtonian ideas of gravity it was Newton's universal law that was applied by NASA engineers and scientists to help land a human on the Moon in July 1969.

Launch of the Saturn V rocket carrying *Apollo 11* to the Moon in July 1969 – all with more than a little help from Newtonian physics (Photo courtesy of NASA)

Famously, the story goes that Newton's gravitational insights came about after observing an apple fall from a tree in the garden at Woolsthorpe, prompting him to think of that same apple-pulling force extending to the Moon and beyond. Although his theory of gravitation didn't emerge as a complete entity during that wonderful year of 1666, the seeds of enquiry were firmly planted in Newton's mind. Acknowledging that his work was not conceived in a vacuum, Newton paid a debt of gratitude to those scientists who had come before him, stating: "If I have been able to see further it was only because I have stood on the shoulders of giants." Those intellectual giants were the likes of Copernicus, Galileo, Kepler, Descartes, Fermat, Boyle, and Hooke, whose earlier work enabled Newton to create a sound mathematical-based approach to physics.

Newton was slow to broadcast his amazing insights into a wide range of phenomena. Soon after being elected a Fellow of the Royal Society in 1672, Newton submitted a paper propounding his views on the composite nature of white light. Far from attracting universal acclaim, the publication prompted assorted grumblings and objections (many of them entirely unfounded) from scientists both at home and abroad. It was Hooke himself, curator of the

Robert Hooke. Newton's antipathy towards Hooke saw that his legacy was expunged from the Royal Society as much as practically possible once Newton had become its president. Sadly, therefore, no contemporary portrait remains of Hooke, but this impression is based on contemporary descriptions of his appearance (Credit: Peter Grego)

Royal Society, who proved to be one of Newton's most vociferous critics, claiming that Newton's experiments with light actually supported Hooke's wave theory more than it did Newton's corpuscular theory. Newton took criticisms of his meticulous work very personally, and a longstanding antipathy between Hooke and Newton began – one of several bitter feuds in which Newton became embroiled over the years. As a result, Newton refused to publish his scientific work for a number of years afterwards.

The Principia

Newton's most prolific period of scientific work took place between 1665 and 1687. In 1684 Newton's friend Edmund Halley (1656–1742) asked him what would be the shape of the orbit of a planet going around the Sun if the force due to gravity was inversely proportional to the square of the distance. Without hesitation Newton replied that the shape would be elliptical, since he had already calculated it. Unable to immediately provide proofs of this, he promised to look through his papers for his calculations. Three years later, with Halley's help, he fulfilled his promise in spectacular style, with the publication of one of the greatest science books of all time – the magnificent *Philosophiæ Naturalis Principia Mathematica* (*Mathematical Principles of Natural Philosophy*). Often referred to as simply the *Principia*, the three-volume work contains explanations of Newton's laws of motion and his law of universal gravitation.

Newton's Three Laws of Motion

Newton's first law basically defines what a force does. A body remains at rest or moves with uniform motion in a straight line unless acted upon by a net external force. It was Galileo who surmised that a ball would travel forever in the absence of frictional forces. He performed experiments with a ball rolling down one slope and climbing up another and found that they reached almost the same height. If friction was reduced to zero he thought the ball would reach the same height, or if allowed to move along a horizontal plane would move forever. A net force acting on a body basically accelerates it either by changing its speed or its direction.

Newton's second law helps define how to measure a force: Force is proportional to the rate of change of linear momentum. Newton's third law states: If body A acts on body B then body B acts on body A with an equal but opposite force. Earth exerts a gravitational force on the Moon, and the Moon exerts a gravitational force on the Earth. Since the Moon has just 1/81 of the mass of Earth the acceleration of the Earth due to the Moon is small. The force on the Moon by Earth would produce a centripetal acceleration that keeps the Moon in its 27.3 day orbit around the Earth. In the case of a rocket, hot gases are ejected from the rocket's nozzle at great velocity, and the rocket is accelerated in the opposite direction. The rate of change of momentum of the gases equals the rate of change of momentum of the rocket.

The Earth–Moon system, imaged in 1998 from the Jupiter-bound *Galileo* space probe (Photo courtesy of NASA)

One famous thought experiment attributed to Newton is called Newton's cannon experiment. He imagined a cannon firing projectiles from a mountaintop; projectiles were fired at ever-increasing speeds in a horizontal direction. Of course, Earth's gravitational force would act on each cannonball, accelerating them towards the center of the Earth. However, the time would come when a cannonball was blasted with such a speed as to continue to fall around the Earth without ever hitting the ground. Its velocity had taken it into a circular orbit. The centripetal force required to keep it in orbit would be provided by Earth's gravitational force.

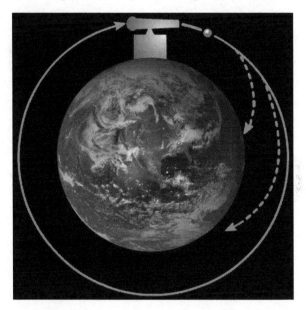

Newton's cannon thought experiment (Credit: Peter Grego)

Newton's Law of Universal Gravitation

According to Newton, the force between any two bodies is proportional to the product of their masses and inversely proportional to the square of their separation distance. This inverse square law of force could explain the acceleration of the Moon to Earth and how all the planets orbited the Sun, as well as the acceleration of an apple to the center of the Earth.

The strength of gravity in the Solar System

Body	Mass in kg	Radius in m	Gravitational field strength "g" in N kg^{-1}	Weight of an 80 kg person in N
Sun	1.99×10^{30}	6.96×10^{8}	274	21,900
Jupiter	1.90×10^{27}	7.14×10^{7}	25	2,000
Earth	5.98×10^{24}	6.37×10^{6}	9.8	780
Mars	6.42×10^{23}	3.39×10^{6}	3.7	300
Moon	7.35×10^{22}	1.74×10^{6}	1.6	130
Ceres	8.7×10^{20}	9.74×10^{5}	0.001	0.08

You would weigh 2.5 times heavier if you could stand on the surface of Jupiter, but your weight on Mars would be less than 40% of your Earth weight. Landing on the dwarf planet Ceres you would feel very strange, as your weight would be reduced by 10,000 times.

Conservation Law of Momentum and Angular Momentum

From Newton's second and third laws we can derive the conservation law of linear momentum. First we can use Newton's third law for a collision between two bodies – in this case the two are interacting and the forces are equal in size, but act in opposite directions, i.e., the sum of the momenta before an interaction or collision is equal to the sum of momenta after a collision.

Gravitational Slingshot

Using Newton's laws the close encounter of a space probe with a planet – an example of an elastic collision – can either accelerate or decelerate a spacecraft by using the planet's gravity and its relative motion to either boost or decrease its velocity. Known as a "gravita-

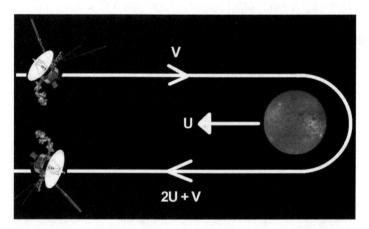

This shows a simple gravitational assist to slingshot a spacecraft around a planet. The spacecraft approaches a planet with velocity v while the planet is moving in the opposite direction at velocity U. The spacecraft is swung around the planet, and its direction is changed by 180°. The net result is that the spacecraft is accelerated and has a final velocity whose magnitude is $V+2U$. Of course, the planet will have its velocity imperceptibly reduced, but the spacecraft has now a speed that has been increased by twice the speed of the planet (Credit: Peter Grego)

tional slingshot" the process sounds like something from a science fiction movie, but it has been used by real spacecraft dispatched to various parts of the Solar System. Both the conservation of momentum and conservation of energy are obeyed in these interactions.

A number of currently operational spacecraft have used this remarkable slingshot effect. The European Space Agency probe Rosetta, launched in 2004, used the gravitational pull of Earth three times and once that of Mars to speed it on its way to Comet 67P Churyumov-Gerasimenko, which it is due to visit in 2014. NASA's MESSENGER probe to Mercury, launched in 2004, will receive a total of six gravity assists during its mission – one from Earth, two from Venus, and three from Mercury. The New Horizons probe, scheduled to reach the distant dwarf planet Pluto in mid-July 2015, received a gravity assist from Jupiter in 2007.

Newton's Cradle

An iconic executive toy that was very popular in the 1980s (and can still be bought for around $20) makes use of the law of conservation of momentum to perform its mesmerizing desktop act. Newton's

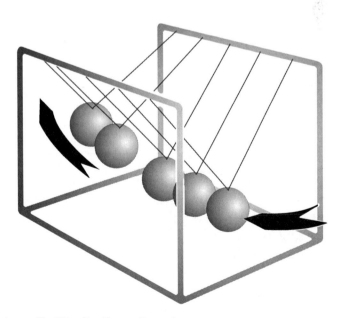

Newton's cradle (Credit: Peter Grego)

cradle usually consists of five steel balls of identical shape and mass, each of which is suspended and free to move in one plane. If N balls are pulled to one side and allowed to fall under gravity, N balls rise at the other end, leaving with nearly the same velocity. Thus both the conservation of momentum and energy are obeyed. Note that in a perfect system we need to assume that the collisions between the spherical steel balls are elastic, which means the total energy is conserved and also the conservation law of momentum is obeyed.

Escape Velocity

If we equate the kinetic energy of a rocket to the potential energy required to escape the gravitational field of a body, such as a planet or a planet's moon, it is possible to derive the escape velocity – the speed that is just sufficient for a rocket to break free of the body's gravitational field. For Earth, escape velocity is about 11.2 km/s or 33 times the speed of sound at ground level.

Escape velocity calculations for various bodies in the Solar System

Escape from	Mass of body (in kg)	Distance from body (in km)	Escape velocity (in km/s)
Sun's surface	1.99×10^{30}	6.96×10^5	621
Jupiter	1.90×10^{27}	7.14×10^4	59.8
Earth	5.98×10^{24}	6.37×10^3	11.2
Escape from the solar system leaving from Earth	1.99×10^{30}	1.50×10^8	42.3
Moon	7.35×10^{22}	1.74×10^3	2.4
Dwarf planet Ceres	8.7×10^{20}	9.74×10^2	0.35
Small 1,000-m diameter asteroid	1.7×10^{13}	1.00×10^0	0.002

You would need to be careful if you landed on a small 1,000-m diameter asteroid, since jumping up at over 2 m/s would enable you to escape its gravitational field.

If the escape velocity from a body is equal to the speed of light $(c \sim 300,000 \text{ km s}^{-1})$, then we have a black hole, from which nothing can escape. For a star like the Sun to be compressed into a black hole its size would have to be reduced from a radius of 665,000 km to less than 3 km. For Earth to become a black hole its size would have to be shrunk to less than 3 cm.

Opticks

The year 1704 saw the publication (in English, rather than Latin) of Newton's *Opticks*, a ground-breaking study of the nature of light and color, including a phenomenon he called the "inflexion" of light (diffraction). In the book, Newton aims "not to explain the properties of light by hypotheses, but to propose and prove them by reason and experiment" – and by so doing, sets forward the basis of all experimental physics. One of the book's greatest achievements was to demonstrate that color – far from being an unpredictable aberration – was a mathematically definable property of light, capable of being observed and measured systematically.

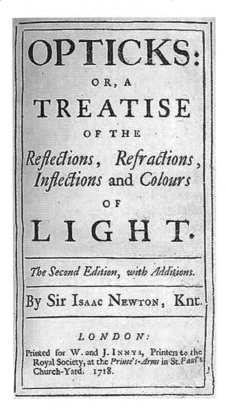

Title page of Newton's *Opticks*

For centuries it had been known that concave glass lenses disperse light by making the rays spread out, while convex lenses bend light to converge into a focus. In the early seventeenth century Lippershey had accidentally discovered that a combination of two different types of lenses could be used as a telescope to collect and focus light, producing a close-up image of distant objects. Early Galilean-type telescopes using a single objective lens to gather light and a simple eyepiece to magnify the image were impaired by chromatic aberration, an effect produced after light passes through a lens and is split up into different colors, ranging from red to violet. These different colors of light – red being the longest and violet the shortest of the visible wavelengths – are unable to be recombined at one focal point by a single lens, producing an image with distinctly colored fringes. "Fatter" lenses with shorter focal lengths display a greater degree of chromatic aberration because they refract light more sharply, producing a greater distance between the focal points of red and violet light. One means by which early telescope makers tried to overcome the problem of chromatic aberration was to use objective lenses with enormous focal lengths.

Using a glass prism – in effect, a small section of a lens – Newton investigated the refraction of light and performed increasingly elaborate experiments that enabled him to discover measurable patterns in light and color phenomena. In a famous experiment, he aimed a beam of sunlight through a hole in a window shutter onto a glass prism, producing a spectrum of colors. He identified seven colors in the spectrum of light – red, orange, yellow, green, blue, indigo, and violet (nowadays, indigo isn't considered a color). By means of another prism – one rotated by 180° – the spectrum could be recombined to produce white light. White light was therefore found to be a mixture of colored rays, each particular color being refracted at a consistent angle by its passage through any particular transparent medium. By passing a narrow band of light of one color through a second prism it was observed that the light remained the same, proving that individual colors of the spectrum could not be further smeared out.

Newton showed that a color's properties were intrinsic and not dependent on the object it scattered from, demolishing notions that white light was pure and that reflection from a body caused it to become a particular color, dependent on the material. Through his work, Newton was first to explain one of nature's most delightful phenomena, the rainbow, as sunlight being split through refraction in spherical raindrops. The different refractive indices of violet to red light in the raindrops – violet light being refracted most, red the least – causes the dispersion of white light into the colors of the rainbow.

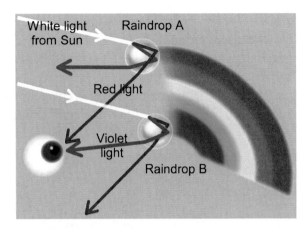

Rainbows produced by dispersion in spherical raindrops. We only see one color from each raindrop. Raindrop A is higher in the sky and directs *red* light towards the observer's eye, while the lower raindrop B directs *violet* light towards the observer (Credit: Peter Grego)

Packets of Light

Some of Newton's work on light was not universally accepted at the time, as it went against some of the prevailing ideas of his day. In particular, Newton's theory that light itself was made of "corpuscles," or particles that traveled in straight lines at a finite velocity, was contradictory to the wave theory of light put forward by the Dutch physicist and astronomer Christian Huygens (1629–1695). In *Traite de la Lumiere* (*Treatise on Light*, 1678) – published as a counter to Newton's work – Huygens put forward the theory that light emanates through space (the "ether") or other media as a self-propagating pulse. Huygens explained the phenomenon of refraction by means of a change in the velocity of light depending on the medium through which it is traveling – faster through space, slower through the air, and even slower through glass.

Newton's corpuscular theory of light explained refraction by using an analogy of balls (light particles) rolling down slopes and encountering different gradients (interfaces between media of different density). When a light particle hits an interface between gradients – say when moving from air to glass – the particle alters course and velocity.

It all seemed pretty logical, and Newton's corpuscular theory of light was to dominate physics for a century or more. But there was one flaw in the corpuscular theory that ultimately proved to be its undoing; the theory demanded that light particles actually travel faster in denser media. In 1850 the French scientist Jean Foucault (1819–1868) conclusively proved the opposite – the speed of light was actually slower in denser media, traveling slower through a lens than it did through the air, just as Huygens theorized.

Inflexions and Rings

Since reading *Micrographia*, Newton had been particularly intrigued by Hooke's descriptions of the colored effects produced in soap bubbles, films of oil on water, and thin sheets of the mineral mica. He was also intrigued by an optical phenomenon that was later to be called "Newton's rings." Now known to be an optical interference pattern, the rings can be produced when monochromatic light (light of one wavelength or color) is shone onto an arrangement of a plano-convex lens placed on top of a large plane glass, creating a series of light and dark fringes starting with a central dark fringe.

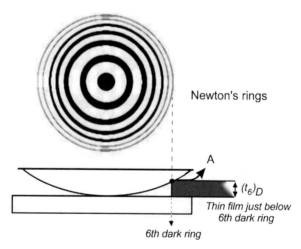

Newton's rings

A

$(t_6)_D$

Thin film just below 6th dark ring

6th dark ring

Newton's rings, an optical interference pattern (Credit: Peter Grego)

The phenomena exhibited by these "inflexions" of light – later to be recognized as interference and diffraction – were not entirely satisfactorily explained by Newton in his corpuscular theory of light.

Newton's somewhat weak explanation of diffraction was that while light appeared to bend around the edges of an object, seeming to act in a wave-like fashion; this was actually a consequence of the scattering effect caused by collisions between corpuscles of light. He explained Newton's rings as being a consequence of light particles undergoing "fits and starts" of easy and difficult reflections.

Newton's rings can be more easily explained by considering light as a wave. If two waves are superimposed the resultant wave is the addition of the two waves, and when in phase they reinforce each other. This is called constructive interference, and it occurs when the optical path difference is a whole number of wavelengths, and likewise destructive interference occurs when the path difference is such that the waves are out of phase.

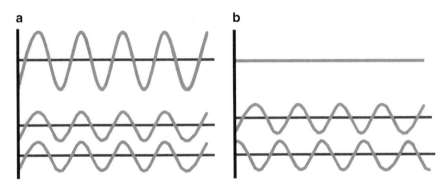

This shows (a) constructive interference and (b) destructive interference (Credit: Peter Grego)

Eventually the wave theory of light came to fore, following work by Thomas Young (1773–1829) in the early nineteenth century that showed diffraction and interference of light through his famous double slit experiments. Later came the great work of James Clerk Maxwell (1831–1879) in his theory of electromagnetic waves, which delivered a resounding blow to the corpuscular theory of light. Maxwell's work proved that visible light, ultraviolet light, and infrared light were all electromagnetic waves of differing frequency; visible light is just a very small portion of the entire electromagnetic spectrum. Using a variety of specialized telescopes today's astronomers examine the Universe in frequencies beyond the visible range; beyond red light ever longer wavelengths of infrared light, then microwave and radio frequencies, are brought into view, while beyond violet light, ultraviolet light, then X-rays and gamma rays reveal their own particular secrets in ever shorter wavelengths.

The story of the nature of light doesn't quite end there. Albert Einstein (1879–1955) showed the particle nature of light in his 1905 paper on the photoelectric effect, but that is another story. Today, light is considered to have a dual nature, with properties of both wave and particle, depending on how it is observed and measured – a consequence of quantum physics.

A New Kind of Telescope

In Newton's day the refracting telescope – essentially unchanged from the time of Galileo in the early seventeenth century – was the visual astronomer's main tool. However, traditional refractors have two major weaknesses: their objective lens introduces an image distortion known as spherical aberration, and the objective also fails to bring all the wavelengths of light to a single focal point, a phenomenon called chromatic aberration, caused by the refractive properties of glass.

Newton realized that chromatic aberration could be overcome if the light gathered by a telescope didn't pass through a lens, but was instead collected and brought to a focus by a mirror with an accurately figured surface. A secondary flat mirror intercepted the cone of light from the main mirror and directed it by 90° to a focus outside the telescope tube, where it could be magnified by an eyepiece.

In 1668 Newton built such a telescope with his own hands and demonstrated it to the Royal Society in 1671. This "demonstration model" had a 3-cm spherical main mirror (made out of highly polished speculum metal) of 15 cm focal length and gave a magnification of 30×. The world's first Newtonian reflector is now in the care of the Royal Society in London and occasionally features in international exhibitions.

Suggestions concerning the use of mirrors in telescopes had been made a number of times during the seventeenth century. Giovanni Sagredo (1571–1620), for example, discussed with Galileo the

A faithful replica of Newton's original reflecting telescope, in London's Science Museum (Credit: Peter Grego)

possibility of using a mirror as a telescope's objective. Scotsman James Gregory (1638–1675) in his book *Optica Promota* (*The Advance of Optics*,1663) suggested the idea of building a reflecting telescope, but he never made a successful instrument of his own design. The Gregorian design was a little more complicated than that of the Newtonian reflector, employing two concave mirrors – a parabolic primary and an ellipsoidal secondary mirror. The primary has a central hole, through which light reflected by the secondary passes to an eyepiece behind the primary. While Gregory was not skilled enough to construct such an instrument, another Scotsman, James Short (1710–1768), successfully capitalized on the Gregorian design during the mid-eighteenth century when he built more than 1,300 instruments, including a 0.5 m telescope (at that time the world's largest telescope).

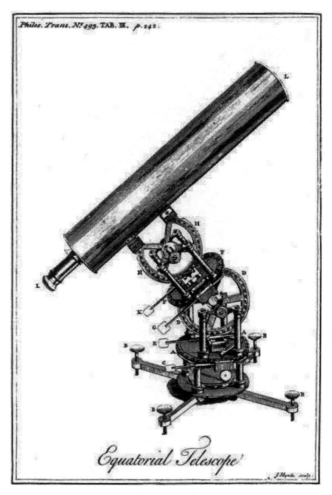

A Gregorian reflector on a tabletop equatorial mount, designed by James Short in 1750

Newton in a Nutshell

Sir Isaac Newton was a forceful character with a complicated, often difficult personality; his main weakness was his inability to deal with public criticism, arguments, or dissension. Most importantly, though, he showed that science should be about the search for truth, and in this respect nobody disputes that Newton was a great man, a towering genius. He transformed the way that science was performed and explained the motions of the heavens. His was an ordered, clockwork Universe, but one which required the hand of a supreme being to keep it working, for Newton was

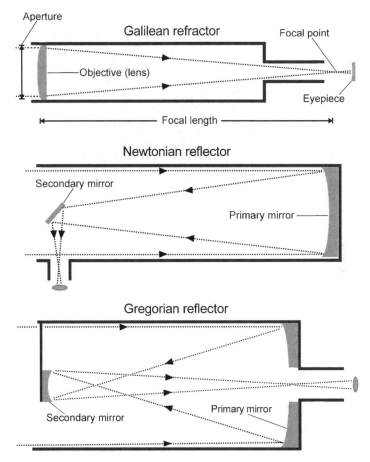

Diagrams of the first telescopes – the Galilean refractor (*top*), the Newtonian reflector (*center*), and the Gregorian reflector (Credit: Peter Grego)

a religious man who thought that if he could glean a small part of how the Universe worked he would begin to understand his creator's handiwork. The great mathematician Pierre Laplace (1749–1827) was enthralled with Newtonian mechanics and felt that any omnipotent being who was aware of every particle's position, mass, energy, and velocity in the Universe would be able to predict its future; just like a clockwork mechanism, the Universe was deterministic.

This account has not touched upon Newton's great contributions to mathematics, nor his more obscure, obsessive interests in alchemy, his research into hidden meanings in the Bible, or his life outside science, which included being Master of the Mint and twice

Peter Grego constructed the primary mirror and mount for his 12-inch Newtonian reflector. (Credit: Peter Grego.)

a member of Parliament. However, in science his works dominated physics and astronomy for more than two centuries. His ideas on absolute space and time were only superceded by Albert Einstein in his Special Theory of Relativity in 1905.

The bedrock of physics and astronomy laid by Newton was built upon by future generations of great astronomers and physicists. As we shall see, astronomers like the Herschels (William, his sister Caroline, and his son John), Friedrich Bessel, Joseph von Fraunhofer, George Hale, and Edwin Hubble were instrumental in transforming optical astronomy in the 200 years after Newton's death. In 1727 – the year that Newton died – astronomers pretty accurately knew the distances within the Solar System out to Saturn. By 1927 astronomers had measured the distances to the nearby stars, measured the abundances of elements in nebulae and stars, had discovered that the Universe was composed of many millions of galaxies, and had found that we were by no means at the center of our own Galaxy, let alone the Universe!

Since Newton's time the known dimensions of the Universe grew from a paltry ten astronomical units (with the stars at an unknown distance) to two billion light years, and we now know that we are living in an expanding Universe. The number of stars in the Universe grew from a few tens of thousands to tens of billions of galaxies, each averaging hundreds of billions of stars. We now know that there are more stars in the visible Universe than all the grains of sand on all the beaches of Earth!

Sir Isaac Newton is quoted as saying: "I do not know what I may appear to the world; but to myself I seem to have been only like a boy playing on the sea-shore, and diverting myself in now and then finding a smoother pebble or a prettier shell than ordinary, whilst the great ocean of truth lay all undiscovered before me." This in a sense was true for astronomy from Newton's time up until the advent of the new astronomies, which then opened up the whole of the electro-magnetic spectrum and added the observations of neutrinos, cosmic rays, and gravitational waves – but we will leave that to Chap. 6.

Let us just finish by stating the results of two polls carried out in 2005, which asked who had been the biggest influence in science: Einstein or Newton? Both polls – one of scientists of the Royal Society and the other a poll of the UK general public by the Royal Society – arrived at the majority view that Sir Isaac Newton had achieved the greatest effect on the history of science!

4. Surveying the Solar System

Lookers and Optic Tubes

During the three centuries following Galileo, most astronomy was carried out with keen eyes glued to the telescope eyepiece, and records of the Sun, Moon, planets, and objects further out in the Universe were made in the form of written notes or observational sketches. Our knowledge of the Universe was improved through the observations of both amateur and professional astronomers, the distinction between which was often more than a little blurred in centuries past. Advances in telescope optics, including the invention of objective lenses that eliminated much of the false color produced by refracting telescopes, inventions of new forms of optical systems, along with engineering improvements, allowed for bigger and better instruments with which to probe the Universe.

In the Netherlands, Christiaan Huygens (1629–1695) constructed lengthy refracting "aerial" telescopes whose lenses were suspended on large frames that were moved by means of ropes and pulleys. Such unwieldy devices, common during this era, were required to overcome chromatic aberration, a defect inherent in simple refractor optics. Chromatic aberration is caused by the inability of a single objective lens to focus all the colors within white light to a single point, making bright objects appear surrounded by unsightly colored fringes; longer focal length objective lenses produce less visually evident degrees of aberration. Huygens also improved eyepiece design by introducing the two-lensed Huygenian ocular – a type of eyepiece still bundled with many modern budget telescopes. Astronomers during this "golden era" of visual telescopic observation managed to ascertain a great deal about the Solar System by simply peering through their telescope eyepieces.

P. Grego and D. Mannion, *Galileo and 400 Years of Telescopic Astronomy*,
Astronomers' Universe, DOI 10.1007/978-1-4419-5592-0_4,
© Springer Science+Business Media, LLC 2010

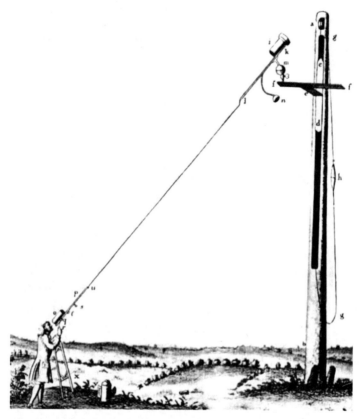

This example of an aerial telescope used by Huygens consisted of two short tubes at each end of a taut rope, one on a universal joint holding the objective lens, the other holding the eyepiece. The lenses were aligned using the rope connecting the two tubes and altitude adjusted using a winch on the main mast

Discoveries on the Moon

The Moon is so large and bright that its surface shows plenty of detail through even the most basic optical equipment – a fact that has delighted lunar observers from Galileo to the present day. It's not surprising that many early telescopic observers chose to study the Moon, to draw its features, and map its surface.

Because the Moon's landscape really did resemble parts of Earth (Galileo had likened it to parts of Bohemia), did it then follow that the Moon had its own atmosphere, and could our

Christiaan Huygens around 1660

satellite support life? Those large dark patches visible with the unaided eye and known as "seas" (Latin, *maria*) were soon found to be relatively flat gray plains devoid of the slightest trickle of water. The absence of any visible bodies of water on the Moon certainly posed a problem for those who wished for some form of lunar life.

Drawing and mapping the Moon – one of the few celestial objects whose surface is clearly and unambiguously presented in the eyepiece of the most modest of telescopes – became a fashionable pursuit among seventeenth-century astronomers armed with their refracting telescopes. Christoph Scheiner (famous for his solar observations, discussed in Chap. 2) published a creditable drawing of the Moon at first quarter phase in his book *Disquisitiones de controversiis et novitatibus astronomicis* (1614).

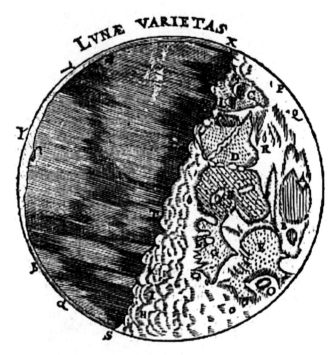

Scheiner's 1614 engraving of the Moon

A few years later the Flemish astronomer Charles Malapert (1581–1630) produced similar drawings based upon single phase studies. In France, the selenographic (Moon mapping) efforts of the Pierre Gassendi (1592–1655) were assisted by the artists Claude Salvat and Claude Mellan. The latter produced three astonishing photograph-like engravings of the Moon in 1636; these depictions are particularly noteworthy because they were produced in Galileo's lifetime, most probably using an instrument employing lenses that had been provided by Galileo.

It's also worth mentioning the lunar work of three Italians: Andrea Argoli's basic Moon studies (1644), Franceso Fontana's stunning depiction of a pockmarked lunar disk (1646), and Eustachio Divini's modern-looking map (1649), each the result of painstaking telescopic observations.

Claude Mellan's beautiful and accurate engraving of the last quarter Moon
(1636) compared with a photograph in 2006 (Credit: Peter Grego)

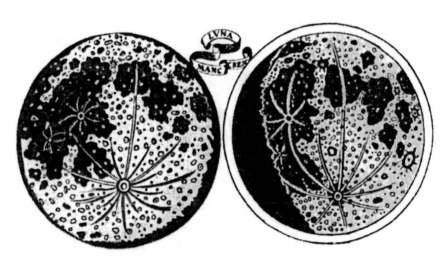

Fontana's pockmarked lunar depiction of 1646

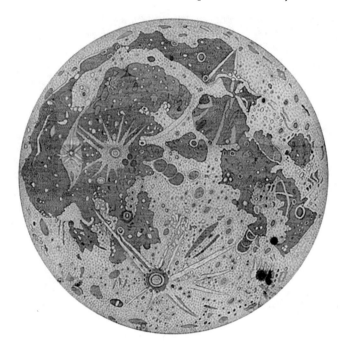

Divini's competent Moon map of 1649

A number of large, detailed lunar maps with their own systems of nomenclature were published in the mid-seventeenth century. The Belgian mathematician and cosmographer to King Philip II of Spain, Michael Florent van Langren (known by the Latinized form of his name, "Langrenus,") (1600–1675), produced a notable work, *Selenographia Langreniana* (1645) in which was included the first map to allocate names to the Moon's features. Unsurprisingly, Langrenus' nomenclature was wholly biased in favor of his patron; the largest single lunar feature (now known as Oceanus Procellarum) was named Oceanus Philippicus in honor of the Spanish king. Most other names on the map proved to be derived from heraldic, aristocratic or biblical sources. Astronomers later threw out the majority of Langrenus's names, but chose to retain the appropriate name for a small area in the Moon's central region, the Sinus Medii. It is surprising that the prominent 132-km crater that Langrenus had the audacity to name after himself escaped subsequent nomenclature revisions.

Langrenus's Moon map of 1645, the first to feature a naming system

Johannes Hewelke (known by the Latinized form of his name, "Hevelius," 1576–1649) published a set of lunar maps and drawings in his *Selenographia* (1647). He personally engraved the illustrations and plates for this work, including three double-page lunar charts of 28.5-cm diameter and 40 descriptions of the Moon's phases. Hevelius's nomenclature was intended to be as neutral as possible by deliberately avoiding naming features after famous philosophers and scientists of the age, for this would have undoubtedly prompted arguments among contemporary selenographers. He named the crater now known as Copernicus after Mount Etna, surrounded naturally enough by a bright area called Sicily; it goes without saying that these features were all placed in the middle of the lunar Mediterranean Sea, now known as Mare Imbrium and Mare Cognitum. Surviving Hevelian names include the mountain ranges of the Alps, Apennines, and the Caucasus Mountains. From his private observatory, Hevelius also made accurate measurements of star positions and produced the splendid *Uranographia* (1687), the most advanced star atlas of its time; it contained 73 constellations, including 12 introduced by Hevelius, of which the latter seven names are still in use today – Canes Venatici, Lacerta, Leo Minor, Lynx, Scutum, Sextans, and Vulpecula.

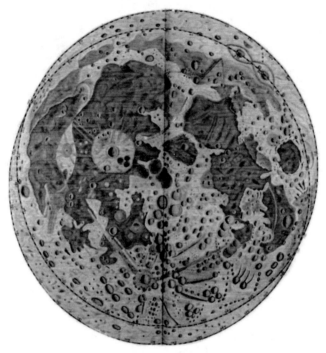

Hevelius's 29-cm diameter Moon map of 1647 (from *Selenographia*)

In 1651 the Italian Jesuit astronomer Giovanni Riccioli (1598–1671) published the important *Almagestum Novum* (*New Almagest*), a book that presented two 28-cm diameter lunar maps based upon the work of Francesco Grimaldi (1618–1663). Riccioli was prudent in his choice of lunar nomenclature. He named most of the lunar seas after moods (such as Mare Tranquillitatis, the Sea of Tranquillity) and the bright highland regions (such as Terra Sterilitatis, the Sterile Land). Craters in the north of the Moon were allocated names of ancient scholars and classical philosophers, surrounded by their pupils and followers; craters in the south were named after prominent renaissance figures. Important

Arabic figures made their way onto the map in a form deemed suitable for western tastes. To name two examples, both of whom were princes and astronomers, the Syrian Ismail Abu'l Fida (1273–1331) was converted to Abulfeda, and the Arabian Muhammed ben Geber al Batani (850–929) was Latinized into the less tongue-twisting (for western tongues) Albategnius.

Few argued with a system that was wise enough to neglect living personalities and which largely avoided blatant bias. However, the theologian Riccioli was no great fan of the heliocentric theory of the Solar System proposed by Copernicus a century earlier. The feature honoring Galileo (a vociferous advocate of the Copernican system) was "cast into the Ocean of Storms." However, it is now thought that the fairly inconspicuous crater in Oceanus Procellarum we now call Galilei is not the same feature indicated by Riccioli; instead, it appears that Galileo was originally represented by the bright swirl feature now called Reiner Gamma.

With the exception of his designations of the highland terrain, Riccioli's nomenclature of more than 200 lunar features is still in use today. It is interesting to note that the first internationally sanctioned system of lunar nomenclature was not agreed upon until 1935 (nearly three centuries after Riccioli's work) by the International Astronomical Union, with the endorsement of a list of 672 names included in *Lunar Formations: Catalogue and Map* by the selenographers Mary Blagg and Karl Müller.

Observatories Come of Age

In the late seventeenth century Giovanni Cassini (1675–1712) helped establish the Paris Observatory as the world's first fully equipped national observatory. He spent much of the 1670s in preparing a detailed 53-cm Moon map based upon his own drawings and those of Leclerc and Patigny. The chart was a monumental effort that showed many lunar features that had been previously overlooked, although somehow the Alpine Valley was not recorded (the first known observation of this conspicuous feature was made by Francesco Bianchini as late as 1728). Only two original prints of Cassini's map are known to exist.

Riccioli's Moon map of 1651

Among the great national observatories set up during the mid- to late-seventeenth century, a royal observatory was established at Greenwich, at that time a small village amid green fields to the east of London. With Britain becoming a burgeoning world power with a booming merchant navy and military fleet, the primary purpose of the observatory was to assist navigators find their way around the globe.

In these days of highly accurate hand-held GPS (global positioning systems) devices that can pinpoint the user's location within a few meters anywhere on Earth, it's easy to overlook the fact that mariners often lost their bearings at sea. While it's fairly straightforward to determine latitude using a sextant aimed at a suitable target in the sky, finding longitude was an altogether more tricky matter in the past. The idea of using the Moon's observed

Giovanni Cassini's magnificent Moon map of 1679, engraved by Claude Mellan

position among the stars to determine longitude was first proposed by Jacques Saint-Pierre (1701–1755).

Before accurate chronometers were devised in the mid-eighteenth century – beginning with the wonderful clockwork mechanisms invented by John Harrison (1693–1776) – mapping the stars and determining the precise path of the Moon through careful observation was therefore of vital national interest and not merely a scientific pursuit. The *Nautical Almanac*, based on Greenwich data, would serve to convey this important information to mariners in a form that would enable them to find their approximate longitude at sea.

King Charles II appointed John Flamsteed as the first "Astronomical Observator" at Greenwich, a title later aggrandized to "Astronomer Royal," whose task was to: "…apply himself, with the most exact care and diligence, to the Rectifying of the Tables of the Motions of the Heavens and the Places of the Fixed Stars, in order to find out the so much desired Longitude at Sea, for the perfecting the Art of Navigation."

Under Flamsteed an extensive and accurate catalog of nearly 3,000 stellar positions was produced, along with a star atlas. During this work, Flamsteed actually recorded the planet Uranus several times, almost a century before its discovery by William Herschel, but failing to recognize its planetary status, he labeled it as the star 34 Tauri.

Peter Grego peers through an original seventeenth-century refracting telescope in the famous Octagon Room of the Old Royal Observatory in Greenwich (Credit: Peter Grego)

In October 1884, an International Meridian Conference organized by U. S. President Chester A. Arthur was held in Washington, D. C., in order to establish an agreed line of terrestrial zero longitude, known as the Prime Meridian. Attended by 41 delegates from 25 nations, the conference selected the Greenwich Meridian (a line running through the center of the observatory's Airy large transit circle) as the official Prime Meridian of Earth, dividing the planet's eastern and western hemispheres. France and Brazil abstained from the vote and for several decades continued to use the Paris Meridian, until confusion caused them to adopt the Greenwich Meridian.

Improvements in Mapping the Moon

In Germany the selenographer Tobias Mayer (1723–1762) introduced the most significant improvement in Moon mapping by
making the first accurate positional measurements of 24 points on
the lunar surface with an accuracy of 1 arcmin in latitude and longitude; these were incorporated into a lunar map, the first chart of
the Moon to possess a distinct system of latitude and longitude.

Tobias Mayer's lunar chart, popular for many years owing to its clarity and
accuracy

Johann Schröter (1745–1816) was inspired to study astronomy after learning of the work of William Herschel, the great
astronomer who discovered Uranus in 1781 (both men were
Hanoverian by birth). His observatory at Lilienthal near Bremen
in Germany was equipped with a number of telescopes, including
two reflectors with mirrors made by Herschel and one made by
Schräder. It is unfortunate that despite having had excellent

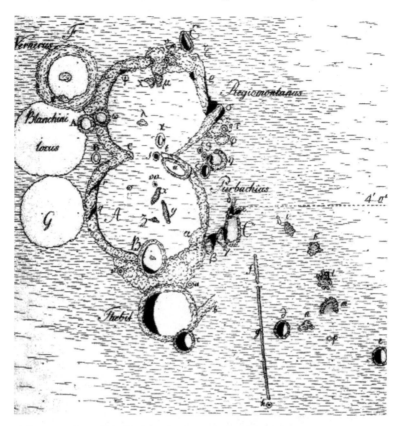

Detail from Schröter's lunar chart of the Purbach region of the Moon, showing the straight Rupes Recta fault

instruments and no end of enthusiasm for lunar and planetary observation, Schröter was disadvantaged to a certain extent by poor drawing skills.

Although he never produced a complete Moon map, Schröter's main contribution to lunar studies was his *Selenotopographische Fragmente* (1791 and 1802), which contained measurements of the positions and heights of the Moon's mountains; numerous charts of lunar rilles were also made. In 1813 French invaders destroyed the observatory buildings, and 34 years of hard-won observational work went up in flames – surely one of the saddest footnotes in astronomical history.

German domination in the field of selenography continued for many decades. Wilhelm Löhrmann (1796–1840) published four sections of a prospective 75-cm map in 1824, a 39-cm lunar map in 1838, and amassed many meticulous observations that went on to be utilized by Julius Schmidt in compiling his 1878 map. Löhrmann accurately determined the positions of 79 craters, chiefly using a 120-mm refractor equipped with a micrometer.

The "dynamic duo" of the golden era of lunar observation was the German astronomer Johann von Mädler (1794–1874) and his wealthy patron Wilhelm Beer (1797–1850). In 1824 the two astronomers commenced a trigonometric survey of the Moon using an excellent 94-mm Fraunhofer refractor. The product of their intensive labors was the *Mappa Selenographica* (1834), a chart 95 cm in diameter and printed in four sections; its structure was based upon micrometric measurements of 105 fundamental points on the lunar surface. Their book *Der Mond* (1837) contained measurements of 148 crater diameters, gave the heights of 830 lunar mountains, and featured a complete description of the Moon's surface. After adding many new names to the face of the Moon, Beer and Mädler introduced the system of lettering

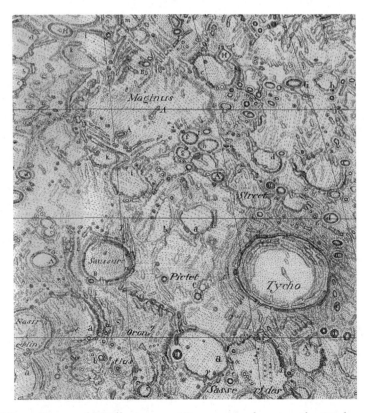

Detail from Beer and Mädler's great Moon map, showing the Tycho region

unnamed craterlets and allocating Greek letters to individual unnamed mountains.

Beer and Mädler considered the Moon to be an essentially dead and changeless world – a proclamation that had an adverse effect upon serious lunar study for decades. If our satellite's features really were, as the astronomers claimed, set rigidly in rock for perpetuity – and if *Mappa Selenographica* had accurately charted most of this detail – then further study could only hope to repeat an already superbly accomplished task. The mid-nineteenth century astronomer might have been under the impression that, after three centuries of lunar mapping, the Moon's surface had finally been captured on paper as well as it could ever be.

Beer and Mädler's mapping efforts weren't superseded until 1879. In that year the German selenographer Julius Schmidt (1825–1884) published a large 195-cm diameter map, in 25 sections, based upon over 40 years of observation and amounting to more than a thousand drawings and many thousands of measurements. This was accompanied by a comprehensive catalog of data. In all, the positions of 32,856 lunar features were recorded (many of which were derived from the works of Löhrmann and Mädler), and more than 3,000 mountain heights, which he had determined using the shadow method; additionally, 278 rilles were charted. Schmidt's map was not without its critics; it was far too cumbersome for amateur field reference, and the original did not even contain the names of features. Because of the huge amounts of data that went towards the work's compilation, many errors found their way into the height measurements, either through misidentification of individual objects or Schmidt mixing up his own calculations.

In the late nineteenth century, valuable British contributions to the subject of selenography arrived in the form of three books, all titled *The Moon*. The first, jointly authored by James Nasmyth (1808–1890) and James Carpenter (1840–1899), was published in 1874. With the use of ingenious models they had constructed, the scientists tried to illustrate the origins of various types of lunar features. For example, cracks in the Moon's crust were depicted in the expansion and contraction of glass spheres covered with dust. It was envisaged that crater formation was dependent upon a once volcanically active Moon. Their famous volcanic fountain

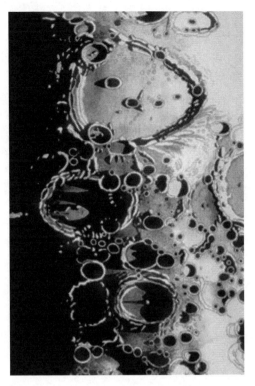

Detail from Schmidt's observational drawing of the Clavius region of the Moon

theory suggested that the outer ramparts of craters were distant accumulations of ash and debris deposited around the site of violent eruption; as activity died down the ejected material landed nearer to the volcanic vent, in most cases ultimately to form a central elevation. Crater ray systems were explained by crustal contraction, fissuring, and in filling with light colored lavas.

Edmund Nevill (alias Neison, 1851–1940) published his Moon book in 1876. It contained a map based upon Mädler's and featured a lively description of the lunar features. Twenty years later Thomas Elger (1838–1897) published a volume that contained a useful map 45 cm in diameter. Both books were welcomed by amateur astronomers.

In 1873 both Nevill and Elger helped found the British Selenographical Society, a group whose aim was the advancement of lunar knowledge. The society disintegrated in 1883 after the death of its president, William Birt, and the resignation of Neison. It didn't take long for the gap in British amateur lunar astronomy to be filled, however, for in 1890 the British Astronomical Association was created, attracting to it many members of the defunct Selenographical Society. Elger became the BAA's first Lunar Section director, and later directors distinguished themselves by producing their own maps of the Moon. In 1910 Walter Goodacre (1856–1938) published a 193-cm diameter map in 25 sections, the co-ordinates for which were based on Samuel Saunder's 1907 catalog of 3,000 photographic positional measurements. Finally, Percy Wilkins (1896–1960) prepared a 2.5-m diameter map, first published in 1946 in 16 sections; impressive for its sheer size and copious detail, the map represents the last great visual map of the Moon.

Percy Wilkins and his 2.5-m map of the Moon (Photo courtesy of Eileen Coombes)

Another great lunar map based upon visual studies was prepared by the prolific German selenographer Philipp Fauth (1867–1943). From his observatory in Landstuhl, Germany, he made intricate studies of the Moon's features using his 160-mm and 175-mm refractors. Despite his obvious observing and drafting skills, Fauth had some highly unusual ideas about the Solar System. He supported the *Welteislehre* (World Ice Doctrine), which held that most of the Solar System's bodies are composed chiefly of ice; the Moon was claimed to possess an icy shell around 200 km thick. Lunar ice had a few notable supporters, none more eminent than the famous nineteenth century British physicist Edward Frankland, who (though he did not believe in *Welteislehre*) maintained that the Moon had once experienced a glacial epoch; he claimed that some of the valleys, rilles, and streaks on the Moon's surface were due to the erosive action of lunar ice sheets in the remote past.

Odd Lunar Happenings

Very early in the history of telescopic observation, those scrutinizing the Moon came to have no doubts that they were looking at a world whose surface is far less dynamic than that of Earth. The dark lunar tracts that some of the first observers had somewhat optimistically called seas turned out to be nothing more than deceptively smooth plains of solidified lava. Much to astronomers' disappointment it was apparent that there were no appreciable expanses of water.

No obvious changes are ever seen on the Moon, other than the predictable shadow play caused by topographic features during the course of the fortnight long lunar day. Visual observers saw that the Moon had no appreciable atmosphere; no winds whipped across its gray surface, no clouds scuttled across its skies, and rain never quenched the dry Moon dust. The nineteenth century science popularizer Richard Proctor once described the Moon on being nothing more than "a dead and useless waste of extinct volcanoes."

However, from time to time astronomers reported (and continue to report) strange activity on the Moon, including colored glows, flashes, and obscurations of surface features, the causes of which may include meteorite impacts and gaseous emissions.

Observational evidence accumulated over centuries shows beyond doubt that the Moon is occasionally host to anomalous short-lived activity known as TLP (transient lunar phenomena). These events seem to occur sporadically in a number of specific small areas of the lunar surface. TLP assume a variety of forms, including isolated flashes or pulses of light, colored glows and obscurations of portions of the lunar surface; some even briefly masquerade as topographic features.

Astronomers have been generally reluctant to accept that our satellite occasionally displays such very obvious signs of activity. Perhaps the lack of widespread recognition of TLP has something to do with the fact that they have been observed mainly by amateur astronomers with limited equipment; few amateurs have been able to secure a photographic record of TLP, let alone employ anything as sophisticated as a spectroscope on these events when they have occurred. The majority of TLP have therefore been inadequately recorded, at least for the purposes of later in-depth scientific analysis.

One of the first earnest attempts to comprehensively list a large number of TLP sightings was made on behalf of NASA by Barbara Middlehurst et al., and published in the *Chronological Catalogue of Reported Lunar Events* (NASA TR R 277). The report gave details of 579 mysterious lunar events dating from November 26, 1540, (pre telescopic) to October 19, 1967. The catalog appeared just a year before Neil Armstrong planted his size 11 boot in the Sea of Tranquillity – strange that such an important and well-funded Moon landing program chose to arm itself with some basic historical TLP data only at the very last minute.

NASA's belated inquiries represented a grudging acknowledgement that the Moon might not actually be the dead world it so convincingly advertises itself to be for most of the time. However, it was in NASA's interest to downplay the idea of an active Moon. Known factors in lunar exploration were hazardous enough to plan for and contend with, without having to admit that there might be some unknown, unpredictable, and uncontrollable threat that might jeopardize the entire $25 billion program.

Anomalies Around Aristarchus

William Herschel noted a distinct red glow in the vicinity of the crater Aristarchus on the evening of May 4, 1783, at a time when that feature was situated on the unilluminated lunar hemisphere. Herschel was observing with a small group at the time, and the presence of the small, faint light was confirmed by others. Through his 22.5-cm reflector the glow appeared to him "as a red star of about the fourth magnitude." On several dates in April 1787 Herschel again recorded prominent TLP, and he became convinced that the lunar surface was experiencing some form of volcanic activity at three separate spots, including the Aristarchus region. He was so convinced, in fact, that on May 20 he invited King George III to view the crater with him using the royal telescope on the grounds of Windsor, though it is not known whether his Majesty became Herschel's observing partner that night.

One of the most notable TLP sightings occurred at 18:45 UT on July 19, 1969, when the crew of *Apollo 11* observed the northwest wall of Aristarchus to be displaying some kind of peculiar luminous activity. At the same time, astronomers at the Institute for Space Research in Bochum, Germany, observing with a 15-cm refractor, noted distinct brightenings in Aristarchus lasting 5–7 s.

The mysterious crater Aristarchus, photographed from the command module of *Apollo* 15 in July 1971 (Photo courtesy of NASA)

It turns out that more telescopes were turned moonward during the period of the Apollo Moon missions (1968–1972) than in the entire 260-year history of telescopic observation preceding it. With such intensive monitoring it is hardly surprising that more anomalous lunar events were reported in this period than at any other time before or since. Though many of these observations could be considered a little dubious, made by inexperienced amateur astronomers keen to note anything that appeared out of the ordinary, some were highly plausible because they were seen by experienced independent observers at different sites.

A handful of dedicated and experienced amateur astronomers across the world continues to monitor the Moon's surface for anomalous activity. Some astronomical societies have implemented their own TLP watch programs; well planned though these projects are, lunar coverage is by no means as complete or as continuous as it could be, so there is ample opportunity for the lunar observer to make a significant contribution to this field of research.

Cooked–Chilled Mercury

Although Mercury is an inferior planet and displays phases just like its neighbor Venus, it is not an easy object to observe, owing to its proximity to the Sun. It is sometimes claimed that Copernicus never actually saw Mercury because of mists arising from the River Vistula in Poland, but this is almost certainly not true, as he had spent a number of years beneath the clearer skies of Italy.

Galileo's telescope wasn't quite powerful enough for him to clearly discern phases on a Mercurian disk which, for the most part, appears between just 5 and 10 arcseconds in diameter. Giovanni Zupi (1590–1650) is usually credited as having been the first to definitely observe Mercury's phases, in 1639.

From time to time at inferior conjunction Mercury moves directly between the Sun and Earth, appearing as a small black spot that slowly moves across the solar disk. Since Kepler's time it has been possible to predict the occurrence of these transits; the first one ever to have been observed took place on November 7, 1631, and was viewed by Pierre Gassendi (1592–1655) from Paris, who wrote about the event in his *Mercurius in sole visus*.

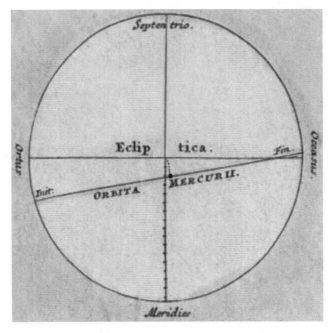

Transit of Mercury observed by Doppelmayer in 1710

Future transits of Mercury

Date	Time (UT, mid-transit)	Duration
2016 May 09	14:57	07 h 30 m
2019 Nov 11	15:20	05 h 29 m
2032 Nov 13	08:58	04 h 26 m

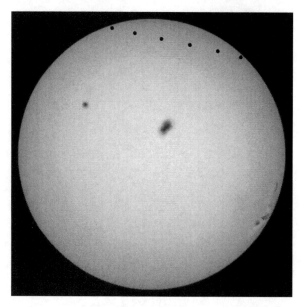

The transit of Mercury on May 7, 2003, observed (Credit: Peter Grego)

It was to be quite a while before telescopes became powerful enough for astronomers to discern markings on the tiny disk of the innermost planet. Subtle Mercurian surface features were recorded between 1780 and 1815 by the great lunar and planetary observer Johann Schröter (1745–1816), who mainly used the 16.5-cm reflector (the mirror of which was made by Herschel) at his observatory at Lilienthal near Bremen in Germany. He consistently observed that the phase of Mercury always appeared more concave than predicted; at predicted dichotomy, when the planet ought to have been a neat half-phase, Mercury seemed slightly crescent shaped. This phase anomaly is caused by the roughness of Mercury's surface and the gradual dimming of light near the terminator; the phenomenon remains observable.

Schröter also noted that the southern horn of the crescent Mercury often appeared somewhat blunted, leading him to speculate that this was caused by the shadow of a huge mountain some 20 km high. Interestingly, the great observer William Herschel found it difficult to discern anything at all on the Mercurian disk, and he discredited Schröter's claim that Mercury had an atmosphere.

Based on his observations, Schröter was of the opinion that Mercury's day was 24 h and 4 min long; Friedrich Bessel used the same observations to arrive at what he thought was an accurate figure of 24 h 53 s and calculated that the axis of Mercury was inclined by a staggering 70° to its orbit around the Sun.

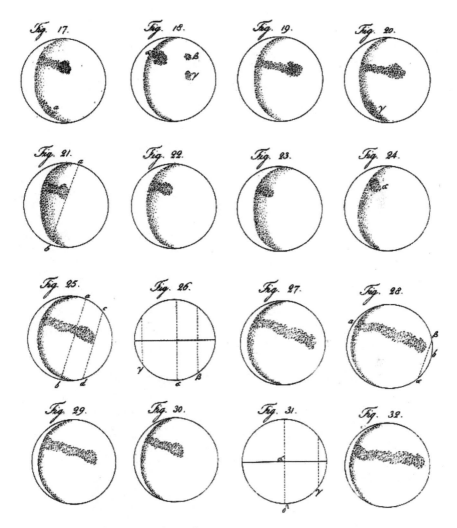

Observations of Mercury by Schröter

Various observers had discerned subtle shadings and brighter spots on Mercury, but little in the way of a concerted effort to understand the planet nearest the Sun took place until the late nineteenth century, when it received the attention of Giovanni Schiaparelli (1835–1910) at the Brera Observatory in Milan, Italy. In 1882 he commenced a series of daylight studies of Mercury (when the planet was high in the sky, above the murk and turbulence of the near-horizon) and succeeded in discerning spots and dusky tracts on the planet on 150 occasions, making drawings of his observations.

Schiaparelli made the case for a thick Mercurian atmosphere – more substantial than that of Mars and more like that of the Earth – noting that the edge of the planet's disk always appeared bright and that there appeared to be a gradual fading of features towards the edges. As for the markings, he wrote:

> The dark spots of the planet, although permanent in form and arrangement, are not always equally apparent, but are some-times more intense and sometimes more faint; and it also happens that some of these markings occasionally become entirely invisible. This I cannot attribute to any more obvious cause than to atmospheric condensations similar to our terrestrial clouds, which prevent more or less completely any view of the true surface of Mercury in any portion.

Schiaparelli used his observations to deduce that Mercury spins on its axis once every 88 days – the same time it takes to complete one orbit of the Sun. Since Mercury has a highly eccentric orbit, it followed that Mercury would display a large libration, where from the planet's surface the Sun would appear to shuffle back and forth along an arc of 47°, creating a broad temperate area in the libration zone between perpetual day and night.

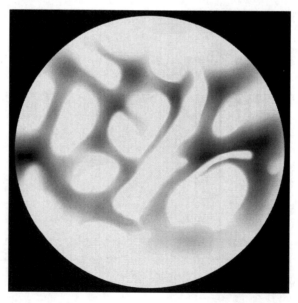

Map of Mercury based on Schiaparelli's 1888 chart (Credit: Peter Grego)

Eugene Antoniadi (1870–1944) was a far more acute visual observer than Schiaparelli, and his visual studies of Mercury using the 83-cm refractor at the Meudon Observatory in France between 1924 and 1929 showed a considerable amount of detail. His book *The Planet Mercury* (1934) contains a detailed map upon which a number of features are given names.

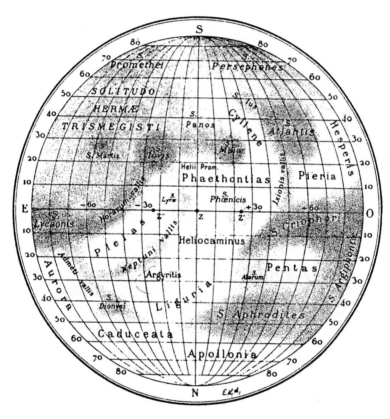

Antoniadi's 1934 map of Mercury

Like Schiaparelli, Antoniadi also thought that Mercury had an atmosphere and that Mercurian dust clouds occasionally diminished the prominence of the planet's permanent surface markings. Antoniadi also thought that Mercury was tidally locked by the Sun's gravity. However, it has since been discovered that Mercury is locked in a 3:2 spin–orbit resonance – it rotates three times on its axis for every two of its orbits around the Sun. This means that alternate hemispheres face the Sun each perihelion. As a result, the planet has two "hot poles" – points on the equator, one at 0° and the other at 180° longitude – that lie directly beneath the Sun when Mercury is at perihelion.

So, why did astronomers suspect that Mercury was in a synchronous 1:1 orbit around the Sun? The answer lies with the fact that they were only observing Mercury and charting its surface features during favorable elongations, when the planet appeared highest in the sky after sunset or before sunrise. Because of this, observers were being presented with a biased view of the planet's surface, a phenomenon termed the "stroboscope effect," whereby an observer is presented with much the same face of Mercury during a number of successive favorable elongations. The stroboscope effect caused observers to wrongly deduce that Mercury was in a captured rotation and that it kept the same face turned towards the Sun throughout its orbit. The stroboscope effect only works over a two or 3-year period. Eventually, orbital dynamics cancel it out, allowing a dedicated observer to view the entire planet over a longer period of time, even if the observations are made only during favorable elongations.

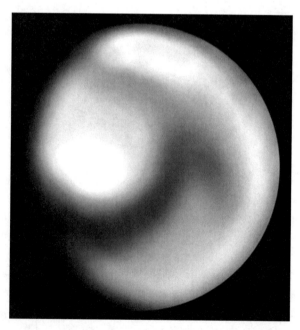

Observation of Mercury on December 9, 2006 (Credit: Peter Grego)

Elongations of Mercury		
Elongation	Date	Distance from Sun (°)
Western	2011 Jan 9	23.2
Eastern	2011 Mar 23	18.2
Western	2011 May 7	26.3
Eastern	2011 Jul 20	26.3
Western	2011 Sep 3	18.2
Eastern	2011 Nov 14	22.2
Western	2011 Dec 23	21.2
Eastern	2012 Mar 5	18.2
Western	2012 Apr 18	27.3
Eastern	2012 Jul 1	25.3
Western	2012 Aug 16	18.2
Eastern	2012 Oct 26	24.2
Western	2012 Dec 4	20.2
Eastern	2013 Feb 16	18.2
Western	2013 Mar 31	27.3
Eastern	2013 Jun 12	24.2
Western	2013 Jul 30	19.2
Eastern	2013 Oct 9	25.3
Western	2013 Nov 18	19.2
Eastern	2014 Jan 31	18.2
Western	2014 Mar 14	27.3
Eastern	2014 May 25	22.2
Western	2014 Jul 12	20.2
Eastern	2014 Sep 21	26.3
Western	2014 Nov 1	18.2
Eastern	2015 Jan 14	18.2
Western	2015 Feb 24	26.3
Eastern	2015 May 7	21.2
Western	2015 Jun 24	22.2
Eastern	2015 Sep 4	27.3
Western	2015 Oct 16	18.2
Eastern	2015 Dec 29	19.2
Western	2016 Feb 7	25.3
Eastern	2016 Apr 18	19.2
Western	2016 Jun 5	24.2
Eastern	2016 Aug 16	27.3
Western	2016 Sep 28	17.2
Eastern	2016 Dec 11	20.2
Western	2017 Jan 19	24.2

(continued)

(continued)

Elongation	Elongations of Mercury	
	Date	Distance from Sun (°)
Eastern	2017 Apr 1	19.2
Western	2017 May 17	25.3
Eastern	2017 Jul 30	27.3
Western	2017 Sep 12	17.2
Eastern	2017 Nov 24	22.2
Western	2018 Jan 1	22.2
Eastern	2018 Mar 15	18.2
Western	2018 Apr 29	27.3
Eastern	2018 Jul 12	26.3
Western	2018 Aug 26	18.2
Eastern	2018 Nov 6	23.2
Western	2018 Dec 15	21.2
Eastern	2019 Feb 27	18.2
Western	2019 Apr 11	27.3
Eastern	2019 Jun 23	25.3
Western	2019 Aug 9	19.2
Eastern	2019 Oct 20	24.2
Western	2019 Nov 28	20.2
Eastern	2020 Feb 10	18.2
Western	2020 Mar 24	27.8
Eastern	2020 Jun 4	23.6
Western	2020 Jul 22	20.1
Eastern	2020 Oct 1	25.8
Western	2020 Nov 10	19.1

Venus the Impenetrable

Galileo's observation that Venus went through a complete cycle of phases was vital in asserting the accuracy of the heliocentric theory of the Solar System. Venus was found to be an Earth-sized world, but its dazzling disk showed few features; those markings which did appear seemed very subtle. Observations of Venus were made during the 1660s by Cassini from Urbino in Italy in an attempt to determine the planet's rotation period. In 1666 he came to the conclusion that the planet either rotated or librated in a period of 23 h, a figure refined to 23 h and 15 min by his son Jacques Cassini (1677–1756) after revisiting his father's observations.

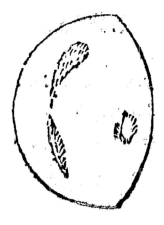

Observations of Venus by Cassini in February and April 1667. Of these he wrote: "Until 28 April, I could not discern any bright spot similar to that which I had seen in February, but on this day, 15 min before sunrise, I re-examined the approximately half-phase disk of the planet and observed a bright spot near the terminator"

Following observations made in 1726–1727, Francesco Bianchini (1662–1729) arrived at a Venusian rotation period of 24 days and 8 h; a century later the seemingly incredibly accurate figure of 23 h, 21 min and 21.934 s was given by Francesco de Vico (1805–1848) following a lengthy series of observations made between 1839 and 1841, many of which were daylight studies.

Schiaparelli didn't detect any obvious Venusian rotation during his observations of 1877–1878, so concluded that the planet's period must be far longer than hitherto proposed. In 1890 he suggested that Venus might be in a synchronous rotation around the Sun (as he had already suggested for Mercury), the planet making one axial rotation in exactly one orbit of 224.7 days. This proposal was by no means accepted in all quarters. From his private observatory at Juvisy-sur-Orge in France, Camille Flammarion (1842–1925) observed Venus extensively and found that his own observations supported a 24-h rotation period. Moreover, his observation of bright caps at the planet's poles convinced him that its axial tilt was not very pronounced.

It appears that all of the above were wrong in their assumptions about Venus's rotation period, and the question wasn't finally settled until radio observations of the solid surface were made during 1964. To everyone's surprise the planet was found to have a retrograde rotation in a period of 243 days; since Venus has a 224.7 day orbital period, its day is 116 terrestrial days long, and the Sun (if it could be seen beneath the planet's thick clouds) rises in the west and sets in the east. Venus's atmosphere, carrying with it temporary cloud features, has its own super-rotation of four terrestrial days, blown along by 100 m/s winds.

Venus's surface is perpetually obscured beneath a thick layer of highly reflective clouds, and speculation about the conditions on Venus's surface was rife during the first part of the twentieth century. Some astronomers imagined that Venus was in a phase of development similar to Earth's Carboniferous Period, with dense swampy jungles and thick forests. Alternatively, it was imagined that Venus was covered with vast oceans steeped in carbon dioxide gas – effectively, seas of foaming soda water.

We now know that Venus is an extremely inhospitable planet. From the perspective of human exploration, it is one of the least inviting places in the entire Solar System. The bright cloud layer preventing its surface from being viewed is stacked between 50 and 70 km above the surface. The clouds themselves are composed of sulfuric acid. At Venus's surface, air pressure is 90 times that of Earth at sea level, and the temperature on all parts of the planet, day and night, is maintained at around 460°C – around 200°C hotter than a kitchen oven at its maximum setting.

Venus' high temperatures are caused by a runaway greenhouse effect. Although much of the solar energy reaching Venus is reflected by its bright cloud tops, some penetrates through the atmosphere and down to the surface, where it is absorbed and then reradiated as heat. Most of the heat emitted by the surface cannot escape from the atmosphere, as it is reflected back down to the surface by the clouds. Venus's searingly high surface temperatures are maintained in this relentless manner, with nothing to break the cycle.

Venus' most prominent dusky features are usually seen near the terminator, from which they extend and fade, sometimes curving towards the poles. A distinct Y-shaped pattern of clouds is sometimes seen spanning the planet's equatorial region from the terminator towards the limb. Dusky collars bordering brighter polar regions are sometimes visible, giving the visual impression of a planet with bright polar caps similar to those on Mars.

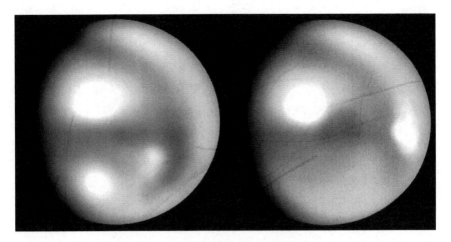

The rotation of Venus's atmosphere, observed February 3, 1996 (Credit: Peter Grego)

Special Effects

Venus's predicted date of dichotomy (half-phase) doesn't always coincide with actual observations. When Venus lies to the east of the Sun dichotomy is sometimes observed to occur some days earlier than the predicted date; at western elongations, observed dichotomy sometimes occurs later than predicted. This phase anomaly is caused by the scattering of sunlight along Venus's terminator; the effects of scattering are more pronounced nearer the planet's edge, where our view is directed through a thicker layer of Venus' atmosphere.

Venus' dark side has occasionally been observed to display a faint illumination when the planet is in a large crescent phase and observed during astronomical night time. Known as the ashen light, the illumination is sometimes patchy and mottled rather than being an homogenous glow. There's little doubt that it is real, rather than an optical illusion, since it has been observed using an occulting bar within the eyepiece to remove the sunlit part of Venus from view. This curious phenomenon has proven difficult to explain, but recent theories as to its cause include the actual glow of Venus's hot surface to lightning flashes within the Venusian atmosphere. Another phenomenon, this time one likely to be illusory, is the anti-ashen light, where during daylight observations the unilluminated hemisphere of Venus appears darker than the background sky.

An exaggerated view of the ashen light faintly illuminating the night side of a crescent Venus (Credit: Peter Grego)

Transits

When Venus moves directly between the Sun and Earth at inferior conjunction, it can be seen as a black circle in transit across the solar disk. Despite covering just one-thousandth the area of the Sun during transit, Venus' apparent angular diameter is so large (around 1 arcmin across) that it can be seen without optical instruments by those with keen eyesight, provided that the eyes are safely shielded from direct sunlight by using a proper solar filter. Special solar eclipse eye shades use a thick layer of aluminized Mylar to prevent most of the Sun's light, heat, and ultraviolet radiation from reaching the eyes by reflecting it away; these can be used for brief but safe views of the Sun. Under no circumstances should such eye shades be worn while observing the Sun through the telescope eyepiece, since the intense magnified energy of the Sun will quickly burn through them and cause permanent eye damage, if not blindness. There are only two safe ways to observe the Sun through a telescope – eyepiece projection onto a shielded white card and whole aperture filtration using a reputable commercially available solar filter.

Transits of Venus take place every 8, 121.5, 8, and 105.5 years. Only seven of these events have taken place since the telescope was invented in the early seventeenth century, the first of these having been observed from England in 1639. The last one took place on June 8, 2004, and it proved to be one of the most widely observed astronomical events in history, viewable in its entirety from the UK, Europe, India, and most of Asia. From eastern North America and much of South America, Venus was already on the Sun's disc as the Sun rose, while the transit was still in progress at sunset for observers in Japan and Australia. There is one more opportunity to view a transit of Venus during the twenty-first century, and it takes place on June 6, 2012.

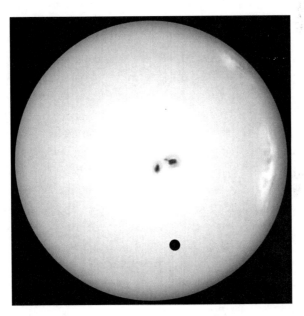

Observation of Venus's transit across the Sun on June 8, 2004 (Credit: Peter Grego)

The circumstances for the transit of 2012 are almost a reversal of those of 2004. Observers in northwestern North America, the western Pacific, northeastern Asia, Japan, eastern Australia, and New Zealand will all have a view of the whole event, from first to last contact. From Hawaii, the Sun is almost directly overhead when the transit begins and ends (before sunset); doubtless there will be a large influx of visitors keen to view the transit from the sunny, friendly islands. The transit is still in progress during sunset from most of North America and northwestern South America. Fortunately, the last part of the transit will be visible as the Sun rises above central Asia, most of western Europe (including the UK), eastern Africa, and eastern Australia. However, Portugal and much of Spain, western Africa, and most of South America will be denied a view of the transit.

First contact occurs when the leading edge of Venus touches the edge of the Sun, at around 22:09 UT (Universal Time). This is not such easy an observation to make as it might first be suspected. The observer will need to do some basic research into where on the Sun's edge Venus is due to make its long awaited entrance. The position angle of Venus's first contact, measured in degrees starting from the northern point of the Sun's disc and moving eastward around the limb, is 40.7°.

Of course, this important information needs to be translated into the orientation of the Sun as viewed through the telescope eyepiece or on the Sun's projected image. Observed through an astronomical telescope with an adequate full aperture solar filter, north is at the bottom of the image and east is on the right. Using a diagonal eyepiece, the image is flipped vertically, so that north is at the top, but east remains on the right. On an image projected from the eyepiece of an astronomical telescope, north is at the bottom, but east is on the left-hand side. Since the orientation of the Sun slowly swings from east to west during the day, its northern point rotating with respect to true north, the orientation of the Sun will vary depending on the time of observation and the observer's geographical location. It is possible for an inexperienced observer to be watching for Venus's ingress not only on the wrong part of the Sun's limb but on the wrong side of the Sun altogether, only to notice the presence of Venus long after first contact, when it has already taken quite a considerable bite out of the northeastern edge of the Sun. Observers using small telescopes and low magnifications will generally notice Venus following its first contact some time – perhaps as much as a minute or so – after observers using high magnifications who are concentrating their attention upon a specific point on the solar limb.

Observers viewing the transit through special solar telescopes with filters tuned to hydrogen alpha frequencies of light may be able to discern the round black disc of Venus a considerable time prior to first contact is seen in white light. This is possible because in hydrogen alpha light, the Sun's hot chromosphere is visible. The chromosphere lies above the photosphere (the body of the Sun visible in normal white light), and glowing prominences loop above the Sun's edge amid a diffuse chromospheric glow. Approaching the Sun, Venus blocks out the hydrogen alpha light from these higher atmospheric features and so may be seen even some time before it makes contact with the Sun's actual chromospheric limb. First contact with the chromosphere precedes first contact with the photosphere by several minutes.

As Venus edges onto the Sun's disc, an unusual phenomenon can be observed. Refraction within Venus's thick atmosphere causes sunlight to be bent around the planet's following limb, producing a narrow arc of light around the edge of Venus. This tiny bow of light beyond the edge of the Sun is fascinating to behold. Indeed, for many observers it represents the most memorable visual highlight of the transit.

Moving at the rapid angular velocity of around 4 arcmin per hour, Venus has fully entered onto the Sun within 18 min of first contact, the point of second contact being the exact moment that the following limb of Venus just touches the edge of the Sun. A phenomenon known as the "black drop" has often been reported in previous transits, causing a degree of uncertainty in individual timings of the moment of second contact. The black drop gives the visual impression of a lingering ligament of darkness extending from Venus's following limb to the Sun's edge, making Venus appear like a drop of black ink suspended from the edge of the Sun. The phenomenon is not a real event, as it is sometimes described in older literature. Instead, it is caused by poor seeing conditions, poor telescope optics, an inaccurate focus, poor observing skills, or a combination of these factors in various measure. Excellent high resolution images of Venus at second contact during the 2004 transit show a perfectly clean contact, with no hint of a black drop effect.

Venus reaches the other side of the Sun, after traversing a solar chord some 25 arcmin long, a little more than 6 h later. Note that because the orientation of the Sun changes gradually as it traverses the sky, the observed track of Venus will appear to be a curved line rather than a straight one. Third contact occurs when the planet's preceding limb touches the Sun's western edge. As the planet leaves the Sun, another arc of refracted sunlight can be traced around Venus's preceding edge, caused in the same manner as that seen after first contact. Fourth contact happens at the very moment when the planet exits the Sun's disc, when the tiny notch disappears and the transit ends for viewers in white light.

Details of the Transit of June 6, 2012

Contact	Time (UT)	Position angle (°)
First contact	22:09:29	40.7
Second contact	22:27:26	38.2
Third contact	04:31:31	292.7
Fourth contact	04:49:27	290.1

The details above have been computed for a theoretical view of the transit from the center of Earth (astronomers call this geocentric data). Actual times will vary slightly depending on the observer's geographic location, but this amounts to no more than a couple of minutes' difference (either earlier or later) at most. Times are given in Universal Time (UT). Mid-transit is at 01:29:28 UT, when Venus is 9 arcmin from the Sun's northern edge (around one-third of the Sun's apparent diameter).

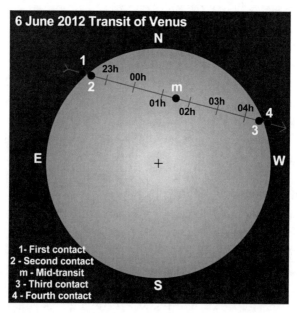

Circumstances of Venus' transit in 2012

The Red Planet

Mars was (and still is) perhaps the most fascinating of all the planets to observe. However, it can only be observed for a few months at intervals of every 2 years or so, when Earth approaches Mars near enough for its disk to grow large enough to observe its features in any kind of detail. Icy polar caps, which grow and shrink with the seasons, crown the Red Planet, and bright clouds sometimes develop in its thin atmosphere. The Martian surface is crossed with light and dark tracts. Unlike the Moon's maria, the markings on Mars vary in intensity and outline from season to season and from one apparition to the next, while retaining their general form and position in the long term. All this adds up to an intriguing world. Astronomers have reasoned that Mars is the most Earth-like of all the planets, perhaps the nearest place in the Universe to harbor some form of life.

Huygens's first observations of Mars, made in October 1659, show dusky features, including a dark V-shaped feature now known as Syrtis Major (a feature familiar to every amateur astronomer who follows Mars through the telescope eyepiece). In 1672 Huygens found that Mars had a bright south polar cap. By observing Mars through a rudimentary micrometer eyepiece he determined the planet's apparent angular diameter; knowing Mars's distance from the Sun he then used simple geometry to work out the true size of Mars, estimating that it was around 60% the size of Earth.

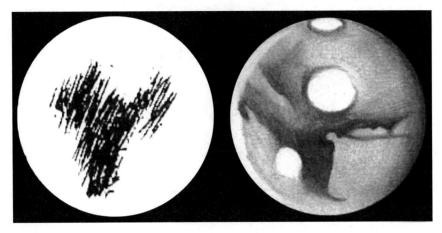

Observation of Mars by Huygens in 1659 (*left*) showing Syrtis Major, compared with a more recent observation of the same area (1989) (Credit: Peter Grego)

Cassini calculated Mars's rotation period to an accuracy of 1.5 min in 1665; a year later he discovered the planet's polar caps. In 1719 Giovanni Maraldi (1709–1788) suggested that these bright polar patches were vast tracts of ice. Mars mapping continued apace through the next century and a half, with notable contributions by Beer and Mädler, Schiaparelli, and Antoniadi.

Are There Martians?

In 1877, Giovanni Schiaparelli claimed to have observed linear markings stretching for thousands of kilometers across Mars. Believing these features likely to be natural geological formations – probably deep faults that had rifted the Martian crust – he named these faint features "canali" (Italian: "channels"). Schiaparelli, a respected astronomer, took care not to speculate that they might in some way be linked with possible life on Mars.

However, Schiaparelli's observations added fuel to a longstanding debate on the possibility that life may exist on Mars. In an era that had witnessed the tremendous engineering feat of

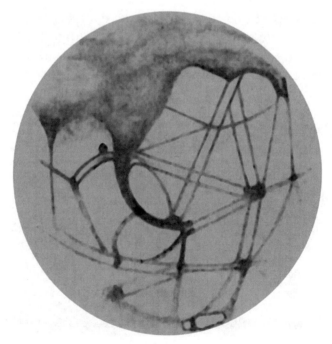

Schiaparelli recorded an extensive network of lines on Mars in this observation of 1888

the Suez Canal, which provided a vital navigable link between the Mediterranean Sea and the Gulf of Suez, the very word "canali" suggested that intelligent Martians might be engaged on similar, though far grander, engineering projects on their own world.

Soon, other astronomers claimed to be able to trace the Martian "canals." Some were not hesitant to proclaim them as evidence of vast waterways hewn out of Mars' deserts by an advanced race of Martians. A wealthy American amateur astronomer named Percival Lowell (1855–1916) was the most vociferous advocate of this theory and went on to write several books on the subject. Lowell claimed that the dusky lines on Mars (and he himself had charted dozens of them) were fertile tracts of vegetation bordering canals that distributed meltwater from the polar icecaps.

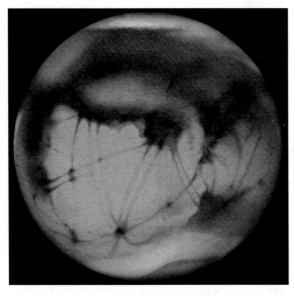

Mars's extensive canal network, based upon maps produced by Lowell (Credit: Peter Grego)

Sadly, intelligent Martians have never existed, nor have the canals of Mars. Many of the observed features were poorly resolved alignments of fine detail, boundaries between areas of different tone, or in some cases optical and psychological illusions.

During the favorable apparition of 1894 eagle-eyed American astronomer Edward Barnard (1857–1923) observed the Tharsis volcanoes and Vallis Marineris using the huge 900-mm refractor of the Lick Observatory. He undoubtedly glimpsed a wealth of topographic detail on the planet's terminator – far too much, in fact, to record accurately. Barnard's exceptional visual acuity is attested to by the fact that many topographic features discovered by space probes are immediately recognizable on his drawings.

Mars's Mini Moons

Two tiny Martian satellites were discovered by Asaph Hall in August 1877, with the 650-mm Clarke refractor of the U. S. Naval Observatory. After much discussion it was decided to name the Martian satellites after the mythological attendants of Mars, namely Phobos (fear) and Deimos (flight). Phobos and Deimos are tiny, city-sized objects, attaining a maximum apparent magnitude (at favorable apparitions) of just 11.6 and 12.8, respectively. Direct visual observation is possible, though it is also seriously hampered by their proximity to the bright Martian disc.

Asteroids, Vermin of the Skies

From the Earth, none of the many thousands of known asteroids are bright enough to be seen with the unaided eye, so it isn't surprising that nearly two centuries had elapsed after the invention of the telescope before the first asteroid was discovered.

The Titius–Bode "Law"

In 1766 the German astronomer Johann Titius (1729–1796) arrived at an apparently simple mathematical formula from which could be derived the distances from the Sun of all the known planets, and which in addition appeared to predict the distances of yet unobserved planets. Titius' formula was published in a book by Johann Bode (1747–1826), and it is now widely known as the Titius–Bode law. According to this formula, not only were there planets further out from the Sun than Saturn, but there was a place between Mars and Jupiter where a planet ought to be located, but had avoided detection because of its faintness.

According to the Titius–Bode law, beginning with 0 for the innermost planet, Mercury, then 3 for Venus; each subsequent planet is double the preceding figure from the Sun: 0 – Mercury; 3 – Venus; 6 – Earth; 12 – Mars; 24 – Asteroid Belt; 48 – Jupiter;

96 – Saturn; 192 – Uranus; 384 – Neptune; 768 – Pluto. If 4 is added to any of these planets and divided by 10, the result is that planet's distance in astronomical units (AU) from the Sun. One AU is the average distance between the Sun and Earth, which is approximately 150 million km. The formula works rather well out to Neptune, whose actual distance falls very short of that predicted. Pluto's actual distance is much closer to this figure.

Planet	T–B law	T–B law distance (AU)	Actual distance (AU)
Mercury	0	0.40	0.39
Venus	3	0.70	0.72
Earth	6	1.00	1.00
Mars	12	1.60	1.52
Asteroid Belt	24	2.80	2.80
Jupiter	48	5.20	5.20
Saturn	96	10.0	9.54
Uranus	192	19.6	19.2
Neptune	384	38.8	30.1
Pluto	768	77.2	39.4

For several years following its publication, the Titius–Bode formula was regarded as a mathematical curiosity. Things changed in 1781, however, when William Herschel discovered Uranus (see below); once its orbit was determined, the new planet was found to fit neatly into the Titius–Bode scheme of things, and the formula became a law overnight.

If the Titius–Bode law was right and it really did have predictive properties, then there ought to be a planet orbiting the Sun between Mars and Jupiter. Convinced that there was such a planet waiting to be discovered, but it was too small and dim to be seen with the unaided eye, a number of astronomers around Europe organized themselves into the "Celestial Police" to hunt it down.

On January 1, 1801 – the very first day of the nineteenth century – Giuseppe Piazzi (1746–1826) discovered the first asteroid, which he later named Ceres (after the Roman goddess of the harvest). Ceres was far too small for its disk to be made out at the telescope eyepiece, but the methodical astronomer charted the object's slow movement among the stars of Taurus. Piazzi thought that the object might be an incoming comet, but after a while it became clear that the new object was a minor planet orbiting the Sun between Mars and Jupiter. Reasoning that there might yet be more minor planets to be discovered, the "Celestial Police" attempted to hunt them down; further success came with the discovery of minor planet Pallas in 1802, Vesta in 1804, and Juno in 1807. To describe the telescopic appearance of these objects, the word "asteroid" (from the Greek, meaning "star-like") was first used by William Herschel. In 2006 Ceres was reclassified from "minor planet" to "dwarf planet" by the International Astronomical Union (see later).

By the end of the nineteenth century several hundred minor planets were known. During the twentieth century, so many asteroids were being discovered photographically that they were sometimes rather unkindly referred to as "the vermin of the skies." Today, the orbits of more than 277,000 asteroids are known.

Jupiter, King of Planets

Astronomers marveled at Jupiter, the largest planet; with its four Galilean moons, it appeared like a miniature Solar System. Bright and dark bands running parallel to the planet's equator had been noted by Zupi in May 1630, and Huygens discovered Jupiter's equatorial bulge – a consequence of Jupiter's rapid spin and its gaseous composition.

In 1664 Robert Hooke (1635–1703) made an interesting series of observations in which he may have been the first to observe Jupiter's Great Red Spot (GRS), a long-lived weather system which still churns up the planet's atmosphere today. At the Paris Observatory in the 1670s Cassini timed the passage of Jovian features at the planet's central meridian and calculated that Jupiter turned on its axis once every 9 h and 55 min.

During the late nineteenth and early twentieth century the prolific British amateur astronomers William Denning (1848–1931) and Stanley Williams (1861–1938) took transit studies to a whole new level by determining the average rotation periods of various belts and zones on Jupiter and noting how features (even in the same belt or zone) appeared to drift in longitude relative to one another over time. Transit timings are still made by amateur astronomers today, since visual estimations can consistently deliver an accuracy within 1° of longitude.

Cassini discovered a delay between predictions of Jupiter's satellite phenomena and timed observations, a phenomenon caused by the finite velocity of light. This allowed his colleague Ole Roemer, in 1675, to calculate the speed of light, arriving at the startling figure of 307,000 km/s (very close to the correct figure of 299,792,458 km/s).

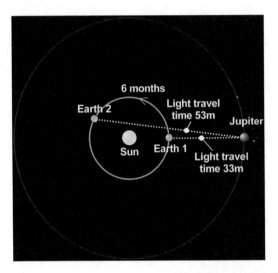

The speed of light was determined in 1675 by noting the delay between predicted and observed timings of Jovian satellite phenomena. Predictions that correctly matched observations at Earth 1 (opposition) were found to be around 20 min late 6 months later at Earth 2, because Jupiter's light had to travel further to the observer (Credit: Peter Grego)

Heinrich Schwabe (1789–1875) made the first definite observation of Jupiter's GRS – or rather, the hollow in the South Equatorial Belt caused by the GRS – in September 1831. Since then the GRS has shifted in longitude, shrunk, and expanded to a certain extent, and slowly changed color from being virtually invisible, through shades of gray and pale pink to a prominent brick red. In 1878 the GRS grabbed Victorian astronomers' attention because of its sheer size and intensity. Measuring some 34° in longitude – an elongated oval around 40,000 km wide and 13,000 km broad – its deep red color was discernable through a small telescope.

Some astronomers seriously thought that the GRS might represent a disturbance among the clouds above a massive volcano erupting onto the planet's unseen solid crust, a vast eddy swirling above the summit of an enormous mountain peak, or a vast raft of material floating upon a turbulent fluid surface. We now know the

GRS to be a long-lived anticyclonic vortex in Jupiter's atmosphere, rotating anticlockwise in a period of about six terrestrial days. It is large enough to easily contain two Earth-sized planets, and despite the fact that it has been slowly shrinking over the past few years astronomers don't know whether future developments will see the GRS begin to grow once more.

In 1939 Jupiter's South Temperate Zone (STZ) became highly disturbed, and three prominent white ovals developed as a result. Designated BC, DE, and FA, these ovals became established in the South Temperate Belt (STB) and lasted for more than 60 years. After passing the GRS and drifting close to each other in 1997, ovals BC and DE collided and merged to form a single larger prominent oval, designated BE. After passing the GRS in late 1999, BE and FA collided in March 2000, creating a single prominent oval, BA. Oval BA has skimmed past the southern edge of the GRS on several occasions without causing much commotion; in addition it turned a distinct red color during 2006, indicating that it is beginning to dredge up exotic compounds from deep within the Jovian cloud layers.

The Ringed Planet

A mystery that had baffled astronomers since Galileo – odd-looking appendages that seemed to cling to Saturn's side and vary in size and shape over the years – was finally solved by Huygens, who explained that Saturn had a flat ring system that nowhere touched the planet. Almost half a century after Galileo had discovered the Solar System's first planetary satellites, Huygens discovered Saturn's largest satellite, Titan, in March 1655.

Cassini correctly suggested that the rings of Saturn consisted of a countless mass of tiny moonlets, all with independent orbits around the planet (Huygens had assumed that the ring was thin but solid). He discovered a gap in the rings – now referred to as Cassini's Division – and between 1671 and 1684 he found four more Saturnian satellites, namely Iapetus, Rhea, Tethys, and Dione.

More structure within the rings was revealed in the following century. A similar, though much narrower, division lies near the outer edge of Ring A, called the Encke Division, after Johann Encke (1791–1865), who observed it in 1837. In 1850, the presence of a broad but faint ring inside the B Ring was noted by William Bond (1789–1859) – a feature whose translucent appearance was described by William Lassell (1799–1880) as resembling "something like a crepe veil covering a part of the sky within the inner ring." Designated Ring C, observers still sometimes refer to this faint feature as the "Crepe Ring."

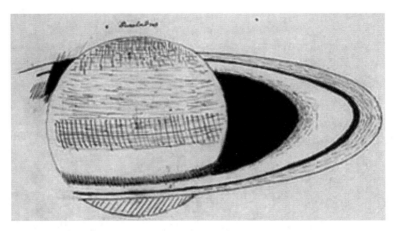

Drawing of Saturn made on September 10, 1851, by William Lassell using his "20 foot telescope"

Numerous visual observers of the Victorian era reported that the main rings were made up of many fainter components, visible only under excellent seeing conditions through large telescopes. In view of the increasing complexity of the rings, along with the undeniable presence of the translucent Crepe Ring, the nature of the rings themselves began to be seriously questioned. In 1856 James Maxwell (1831–1879) asserted that the rings could not possibly be solid; any such feature would be immediately torn up by Saturn's gravity, as the inner parts of the rings attempted to revolve around the planet faster than the outer parts. In fact, the rings lie entirely within a zone of gravitational disruption known as the Roche Limit – the critical distance from a planet at which a sizeable solid body is gravitationally disrupted and torn apart by tidal forces. Maxwell realized that the rings must comprise "an indefinite number of unconnected particles."

Proof of the particulate nature of Saturn's rings came in the form of spectroscopic observations made by James Keeler in 1895;

his measurements of the Doppler shifts of various parts of the rings proved that their orbital periods increased with distance from Saturn. Each of the countless billions of constituent particles within the rings – from dust grain-sized motes to house-sized boulders – behaves as a tiny independent satellite, obeying Kepler's laws of planetary motion.

Saturnian Spots

Despite its apparent blandness, Saturn's atmosphere occasionally plays host to large white spots that suddenly well up and dominate the globe for several months. White spots were observed in 1876, 1903, 1933, 1960, and most recently in 1990 – appearing about every 30 years or so, when Saturn's northern hemisphere has been tilted strongly towards the Sun (note that Saturn's orbital period is 29.46 years).

American amateur astronomer Stuart Wilber discovered the 1990 white spot on September 24 of that year using a 25-cm reflector. In what was to become the most spectacular Saturnian atmospheric upheaval since 1933, the spot enlarged into a broad oval larger than Jupiter's Great Red Spot and was easily visible through small telescopes. After several weeks, 1,700 km/h jet streams in Saturn's upper atmosphere began to alter the spot's appearance; as it grew in longitude, it developed an extensive trailing area that eventually encircled the planet at its latitude. The newly operating Hubble Space Telescope was rescheduled to make observations of the storm; its high-resolution images showed considerable detail, notably scalloping and festoons, similar to those dark features that appear on the southern edge of the North Equatorial Belt of Jupiter. By early November, the spot had been spread into a broad, fairly homogenous bright belt that persisted into the following apparition.

Titan

Titan, Saturn's biggest moon, is large enough to be discerned as a small disk through large telescopes, given good seeing conditions. In 1952 attempts were made by Lyot, Camichel, Dollfus, and Gentili to resolve surface detail, using the 57.6-cm refractor of the Pic du Midi Observatory in the French Pyrenees. The results seemed contradictory. A light central belt was at times seen at varying angles to Titan's equator; a dark equatorial belt was occasionally observed, with a light north polar region. At the time of the observations Titan was known to have an appreciable atmosphere because Gerard Kuiper had spectroscopically detected an extensive veil of methane gas in 1944.

Herschel Discovers Uranus

Uranus was discovered on March 13, 1781, by William Herschel (1738–1822) during one of his routine sweeps of the sky using his 15.5-cm self-made reflector. Then in Taurus, Uranus appeared as a star of magnitude 5.6, but its tiny, clearly discernable disc gave the astronomer cause to suspect that the new object might be a comet. Herschel reported:

> The power I had on when I first saw the comet was 227. From experience I know that the diameters of the fixed stars are not proportionally magnified with higher powers, as planets are; therefore I now put the powers at 460 and 932, and found that the diameter of the comet increased in proportion to the power, as it ought to be, on the supposition of its not being a fixed star, while the diameters of the stars to which I compared it were not increased in the same ratio. Moreover, the comet being magnified much beyond what its light would admit of, appeared hazy and ill-defined with these great powers, while the stars preserved that lustre and distinctness which from many thousand observations I knew they would retain. The sequel has shown that my surmises were well-founded, this proving to be the Comet we have lately observed.

After a number of careful positional observations had been made during the following weeks and months, and its path charted, it became clear that the new object lay far beyond Saturn. It was found to have a decidedly non-cometary, near-circular orbit, and there was no doubt that Uranus was a bone-fide planet – and a very large one at that.

Realizing the importance of the discovery, King George III took Herschel under his wing and became his patron, granting an annual stipend of £200 provided that he relocate from Bath to Slough near

Grego using a replica of the telescope that William Herschel used to discover Uranus from his garden in Bath, England (Credit: Peter Grego)

William Herschel about to conduct an observing session in his garden in Bath, England (Credit: Peter Grego)

William Herschel's "Great 40-foot Telescope," which was erected at
Observatory House in Slough, England. With this instrument, which had
a 120-cm mirror, Herschel discovered Enceladus and Mimas, the sixth and
seventh moons of Saturn

Windsor Castle in order that the royal court might enjoy views
through Herschel's telescopes. Accepting the offer, Herschel decided
to name the object *Georgium Sidus* (George's Star) in honor of his
new patron – a decision that proved highly unpalatable in Europe
and America. We have the German astronomer Johann Bode to
thank for suggesting the name Uranus (mythologically speaking,
Uranus is Saturn's father, while Saturn is Jupiter's father). It wasn't
until 1850 that the HM Nautical Almanac Office finally dispensed
with using *Georgium Sidus* and switched to Uranus.

Uranus' discovery almost doubled the known size of the
Solar System. The planet orbits the Sun at an average distance of
2.9 billion km (more than 19 times the distance between Earth and
the Sun) in a period of 84.3 years. With an equatorial diameter of
51,118 km, gas giant Uranus has a mass of 14.5 Earths and a volume
of 63 times that of Earth.

Last Planetary Outpost

Careful observations of Uranus' position in the decades following its discovery suggested that its orbital path was being gravitationally perturbed by a large unknown planet lying further out in space. In 1846, the gifted mathematician and astronomer Urbain Leverrier (1811–1877) calculated the probable position of this mystery planet and communicated his predictions to Johann Galle (1812–1910) of the Berlin Observatory. Wasting no time in beginning his search, Galle discovered Neptune on September 23, 1846, using the observatory's 22.5-cm refractor; the planet was within just 1° of its position as predicted by Leverrier. It shone at magnitude 7.8 and measured 2.3 arcminutes in apparent diameter and was therefore too small and faint to be seen without optical aid.

As with Uranus, a certain amount of controversy accompanied the naming of the newfound planet. Claiming its discovery, Le Verrier chose to name it Neptune after the Roman god of the sea, but later opted to name the planet after himself. Outside of France, this proved an unpopular suggestion, and Le Verrier's original suggestion eventually held sway.

Neptune orbits the Sun in a near-circular path at an average distance of 4.5 billion km – 30 times the distance of Earth from the Sun – so far that it takes 164.8 years to orbit the Sun. On May 29, 2011, the planet will have completed exactly one orbit since its discovery. Neptune is a gas giant measuring 49,528 km at the equator, making it slightly smaller than Uranus. The presence of methane in its atmosphere – an effective absorber of red and yellow light – gives the planet a distinctly blue hue.

Distant Ice Worlds

Searches for a planet beyond Neptune were begun in the early twentieth century after some astronomers claimed that such an object was responsible for causing perturbations in the orbit of Neptune – just as Uranus' orbit is perturbed by Neptune itself. A photographic search for the Solar System's ninth planet began at Lowell Observatory in 1927; success came on February 18, 1930, when faint images of a tiny moving speck of light were discovered on photographic plates taken by Clyde Tombaugh using a 330-mm astrographic telescope. Jubilation turned to slight disappointment, as it soon became evident that Pluto was far too small to be responsible for perturbing Neptune; Pluto's discovery turns out to be one of those curious coincidences that have cropped up from time to time in the history of astronomy.

Pluto circles the Sun in a period of 248.5 years, in a far more eccentric and highly inclined orbit than any of the major planets, ranging from a 4.4 billion to 7.5 billion km from the Sun (the latter being more than 50 times the distance between Earth and the Sun). If viewed from above, it would appear that the orbits of Pluto and Neptune would actually cross each other; however, since their orbital planes are considerably inclined to each other, there is no danger of a planetary collision at any point in the future.

Pluto was once officially credited with being the Solar System's ninth planet, even though it was found to be just 2,274 km across, making it smaller than seven planetary satellites, including our own Moon. Pluto's status as a planet seemed to be assured when it was found to have a large satellite in 1978. Measuring 1,172 km across, the satellite, named Charon, is more than half the diameter of Pluto. The pair could even be considered a double planet, as they both revolve around a common center of gravity in space. In 2005 the Hubble Space Telescope was used to discover two more moons around Pluto – tiny chunks of rock and ice called Nix and Hydra.

However, in the early twenty-first century astronomers began to question Pluto's planetary status after the discovery of a number of sizable icy worlds scattered around the outer Solar System. In 2005 a team at the Palomar Observatory led by Mike Brown discovered an object now known as Eris; with a diameter of 2,500 km and 27% more massive than Pluto, Eris was originally claimed to be the Solar System's tenth planet. Like Pluto, Eris is a trans-Neptunian object with an eccentric, highly inclined orbit, its distance from the Sun ranging between 5.7 and 14.6 billion km and a period of 557 years. With its small satellite, Dysnomia, the pair is the most distant known planetary body in the Solar System, currently three times the distance of Pluto. Eris won't be at its closest to the Sun until 2256, and in 800 years' time it will actually be closer to the Sun than Pluto for a while.

Hubble Space Telescope view of the dwarf planet Pluto and its satellites (Photo courtesy of NASA/HST)

Other large trans-Neptunian objects include worlds with strange names such as Makemake, Sedna, Varuna, Haumea, Orcus, and Quaoar, each of which is smaller than Pluto; the latter three have their own small satellites.

Because Eris is larger than Pluto, its discoverers and NASA were tempted to call it the Solar System's tenth planet. In 2006 the International Astronomical Union (IAU) was moved to define the term "planet" for the first time. As a result, distant Pluto, Eris, Haumea and Makemake, along with Ceres in the main Asteroid Belt, were deemed to be "dwarf planets."

Comets, Celestial Ghosts

With their sudden and unexpected appearance in the night skies, their motion across the heavens, and their changing appearance from night to night, comets tended to strike awe and fear into our ancestors in centuries gone by.

We now know that comets consist of a relatively tiny solid nucleus, typically a dozen km across, made up of silicate rock and ices – mainly water ice, but also various proportions of ammonia, methane, nitrogen, carbon monoxide, and carbon dioxide ice. Their composition has led to the somewhat ignominious description of the "dirty snowball." Most comets are thought to have been formed out of the original interstellar material that condensed to form the Sun and the Solar System, although a few may have been gravitationally purloined from other nearby star systems.

When a cometary nucleus approaches the inner Solar System, it begins to react to increasing levels of solar energy. Heated by the Sun, the ices on the surface of the nucleus sublimate, changing from a solid state to a vapor (gas). These gases ionize as they pick up electrical charge from the solar wind, forming a large cloud around the nucleus known as a coma, which can grow to as much as 100,000 km in diameter. Pressure from the solar wind pushes the ionized gases directly away from the Sun, and a long, straight ion tail is formed.

Rather than sublimating evenly across the comet's sunlit face, there are usually a number of more active regions, producing jets that throw out small dust grains that were originally embedded in the icy crust. Pressure from the solar wind causes the released dust grains to lag behind the nucleus, producing a separate dust tail that develops alongside the gaseous ion tail. These dust particles provide the stuff of meteor streams.

Comets vary enormously in the amount of dust and gas that they produce. Some comets are more dusty than others; some display just a broad fuzzy globular coma, while others may develop a magnificently structured coma, plus bright ion and dust tails. Some of the more active comets become very prominent in the night skies. Nobody who saw Comet Hale-Bopp during 1997 is likely to forget what a magnificent spectacle it made, with its bright star-like nuclear condensation, curved yellow dust tail, and straight bluish ion tail. Other comets – indeed, the majority of comets observed – never reach naked-eye visibility and remain dim objects seen only through binoculars and telescopes.

New comets fall into the Solar System from virtually any direction, but no comet has been observed to have an orbit that suggests an interstellar origin, leading astronomers to the conclusion that they originate far beyond the planets, in realms so distant that the Sun appears as a bright star-like point. This zone,

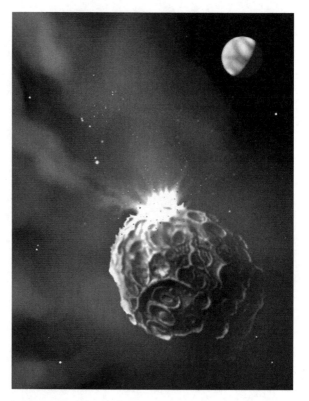

Impression of the nucleus of Halley's Comet during a close encounter with Venus (Credit: Peter Grego)

populated by countless icy cometary nuclei, is known as the Oort Cloud and consists of a vast shell that extends a quarter of the way to the nearest stars. The Oort Cloud is the source of all long-period comets.

Hundreds of comets have well-known orbits lying relatively close to the Sun, within planetary realms; these periodic comets make regular, predictable visits to the inner Solar System. Although most of them are too faint to observe through average backyard telescopes even when they are at their brightest near perihelion, a number of them are familiar to amateur astronomers. Halley's Comet, for example, has a 76-year orbital period, and at its furthest lies beyond the orbit of Neptune. With an orbit of just 3.3 years, Comet Encke has the shortest known period of any comet, its orbital path taking it from inside the orbit of Jupiter to inside Mercury's orbit.

There are 164 periodic comets currently designated a number on a list maintained by the Minor Planet Center and prefixed with P (for periodic), in order of the discovery of their periodic nature. Halley's Comet is referred to as 1P Halley. In addition, there are more than 160 periodic comets without a designated number and an additional several dozen asteroids that display all the attributes of cometary nuclei but which don't approach close enough to the Sun for their ices to sublimate.

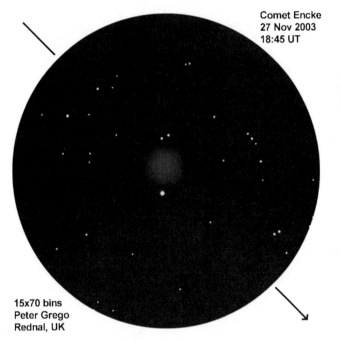

Comet Encke
27 Nov 2003
18:45 UT

15x70 bins
Peter Grego
Rednal, UK

Binocular observation of Encke's Comet, a comet with the shortest known period (Credit: Peter Grego)

5. A Bigger Picture Unfolds

Galileo's small telescope showed that the Milky Way was composed of a multitude of stars that were too faint to be resolved individually with the unaided eye. Unlike the planets, which when magnified through a telescope displayed a small but distinct disk, the stars remained as point-like objects. This, along with the fact that the stars displayed no discernable parallax through the instruments in use in the seventeenth century, suggested that the stars were at a truly unimaginable distance from the Solar System. French philosopher Blaise Pascal (1623–1662) conveys a vivid sense of the human reaction to this new-found scale of the Universe when in his *Pensées* (*Thoughts*, 1670) he writes of a man without God: "engulfed in the infinite immensity of spaces of which I am ignorant and which know me not… The eternal silence of these infinite spaces frightens me."

Pascal had some reason to be fearful about the wider Universe. During the 1640s he had conducted experiments in barometry; by demonstrating that atmospheric pressure decreased with height, his work supported that of Evagelista Torricelli (1608–1647), who proposed that Earth's atmosphere did not extend indefinitely, and that beyond it lay the harsh vacuum of space. After a visit to Pascal in September 1647, which turned sour, philosopher René Descartes (1596–1650) wrote to Huygens and commented that "Pascal has too much vacuum in his head."

Descartes imagined that the Universe was composed of a "subtle matter" termed "plenum," which swirled around objects in whirlpool-like vortices, chivvying comets, satellites, and planets along in their orbits.

P. Grego and D. Mannion, *Galileo and 400 Years of Telescopic Astronomy*,
Astronomers' Universe, DOI 10.1007/978-1-4419-5592-0_5,
© Springer Science+Business Media, LLC 2010

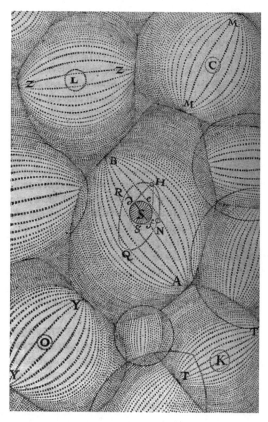

Descartes' notion of a plenum-powered Universe of vortices, from *Principia philosophiae* (1644)

Descartes believed that the stars were Sun-like objects, each with its own Solar System, but thought that the Universe had a finite number of stars. Still, it was difficult to imagine a finite Universe – if it had a boundary, what lay beyond?

It followed that if the stars themselves were like the Sun – but so far away that they appeared as mere points of light – then perhaps the Sun itself didn't lie at the very center of Universe, but was just one of a broader mass of stars contained within a vast star system containing countless thousands of stars. In the mid-eighteenth century, Thomas Wright was most perceptive in suggesting that the band of the Milky Way was caused by our view from deep within a vast, flat, millwheel-shaped star system; he found no evidence that the Sun lay at the Galaxy's center, or that the system was finite in breadth.

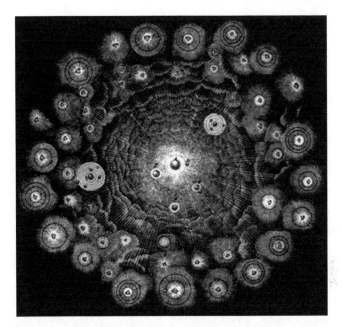

Bernard Fontenelle published this view of a galaxy full of extrasolar systems in his *Entretriens sur la pluralite des mondes* (1682)

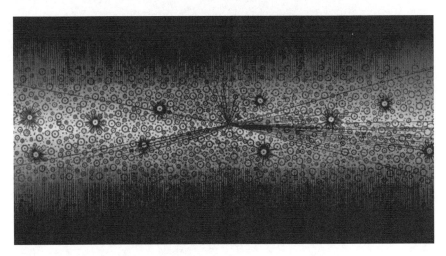

Thomas Wright's 1750 concept of the Galaxy's layout

Herschel's Insights

From southern England, using his own home-made telescopes, William Herschel conducted a thorough "review of the heavens" during the late eighteenth century. He succeeded in producing a catalog of star clusters and nebulae that surpassed the relatively modest work of Charles Messier. Herschel noted the relative motion of stars, proving that they were not simply points of light fixed on some distant heavenly vault. Some stars were shown to be physically associated with one another – stars with common orbits around each other. Herschel went on to discover the actual motion of the Sun and the Solar System through space, a finding that further removed the Sun from the center of the Universe.

Herschel applied the technique of star counting during his telescopic sky surveys, in which he counted the number of stars visible within any particular field of view and noted their apparent brightness. Assuming that the stars were all more or less the same actual brightness and spread through an equal volume of space in different parts of the Galaxy, Herschel gained an idea about the actual shape of the Milky Way Galaxy. He was unable to come to a definite conclusion whether the Galaxy was a flattened disk of infinite diameter, or whether it was finite and bounded on all sides by empty space.

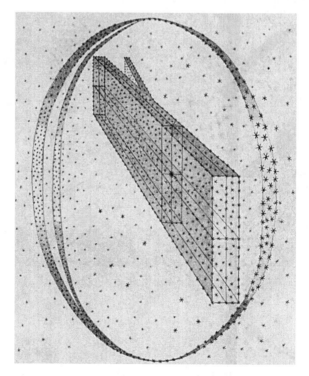

Herschel's 1784 speculation on the structure of the Milky Way, showing (*center*) a representation of the layout of the Galaxy in the vicinity of the Sun and (*outside*) how the Galaxy might appear from a great distance

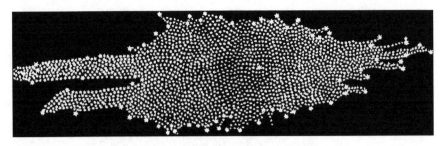

Herschel's speculative view of a cross-section through the Milky Way; the Sun is estimated to lie just right of center

Charting the Stars

A catalog compiled by the Greek astronomer Ptolemy in 150 CE featured 48 constellations, including the 12 constellations of the zodiac. Many of these constellations had been in use for thousands of years previously. Constellations are a handy means of identifying and locating objects in the sky. There are 88 constellations of various shapes and sizes officially recognized today, all contained within neatly defined boundaries, making the celestial sphere look like an enormous 3D jigsaw puzzle. In addition to Ptolemy's original 48, more recent additional constellations include those covering the southern skies.

Look at any star map published between, say, 1600 and 1800, and it may at first be difficult to distinguish the stars from the beautifully engraved depictions of mythical figures cavorting about the heavens; close scrutiny will reveal the stars as multi-pointed jewels of varying sizes, according to their brightness. Though they may look like beautiful works of art rather than scientific documents, the star maps produced by astronomers such as Hevelius in the mid-seventeenth century and John Bevis (in the mid-eighteenth century) were depicted with the best accuracy then attainable, each star's position having been carefully noted at the telescope eyepiece. Being equipped with accurate charts of the heavens enabled astronomers to identify new planets, comets, novae, and deep sky objects, and it was possible to accurately plot the motion of the Moon and planets among the stars.

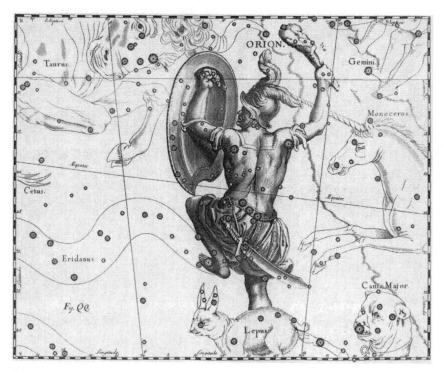

The constellation of Orion according to Hevelius in his *Uranographia* (1690). Note the east-west flipped view, as it would appear on the outside of a celestial globe

Delving into Deep Skies

Permanent faint patches of light among the "fixed stars" have been familiar to watchers of the skies since ancient times, the most obvious being the patchy luminous band of the Milky Way that circumscribes the celestial globe. More compact patches known as "nebulae" (Latin: "clouds") visible with the unaided eye include the Large and Small Magellanic Clouds in the southern hemisphere, the star cluster Praesepe, the double cluster of h and Chi Persei, the Andromeda Galaxy, and the Great Nebula in Orion, none of which can be resolved into stars or seen in any great detail without optical aid.

Hodierna's Nebulae

As soon as the telescope began to sweep the night skies astronomers began to discern detail within the known nebulae, in addition to making discoveries of entirely new nebulae. We've seen that Galileo noted how the Milky Way and Praesepe appeared to be composed of a multitude of stars too faint to be individually seen with the naked eye, and that the star cluster of the Pleiades contained dozens more stars than had previously been known. Galileo's contemporary, the German astronomer Simon Marius, achieved a notable first by "rediscovering" the Andromeda Galaxy in 1611 and describing it as like a "flame seen through horn."

In 1654 a list of forty nebulae, along with finder charts and observational drawings, was published by the Sicilian astronomer Giovanni Hodierna (1597–1660), part of a larger project to map the entire sky in 100 sections that was never completed. Of the 40 deep sky objects in the list, about a dozen are simply asterisms (small groupings of stars) or are not described well enough to be identified with any certainty. Hodierna classified nebulae into three types: *Luminosae* such as the Pleiades, in which some stars are visible with the unaided eye; *Nebulosae* such as Praesepe, which resolved into stars through a telescope; and *Occultae* such as the Andromeda Galaxy, which appeared as unresolved patches through a telescope.

Hodierna's 1654 depiction of the Orion Nebula, the first known drawing of this well known deep sky treasure. The three stars within the nebula are likely to be Theta1, Theta2 A, and Theta2 B Orionis

Hodierna's work was not well known at the time and has only recently been rediscovered. It shows that a keen interest was being taken in deep space at a time when the Solar System dominated much astronomical debate.

Luminosae	Modern Designation
HL1	M45 (Pleiades)
HL2	Hyades (Melotte 25)
HL3	Coma Berenices star cluster (Melotte 111)
HL4	Alpha Persei moving Cluster (Melotte 20)
HL5	M42
HL6	Asterism including Lambda, Phi1, and Phi2 Orionis
HL7	NGC 6231
HL8	Asterism in Aquarius
Nebulosae	Modern Designation
HN1	M44 (Praesepe)
HN2	M7
HN3	NGC 869/884 (h and Chi Persei)
HN4	M6
HN5	Asterism of Nu1, Nu2 Sagittarii
HN6	M8 (Lagoon Nebula)
HN7	M36
HN8	M37
HN8	M38
HN10	"Brocchi's Cluster" (Collinder 399)
HN11	Asterism around 88 Herculis
HN12	Asterism in front of the head of Capricornus
Occultae	Modern Designation
HO1	Asterism in Coma Berenices
HO2	Asterism in Coma Berenices
HO3	M31 (Andromeda galaxy)

Messier's List

One of visual astronomy's greatest comet hunters, French astronomer Charles Messier (1730–1817), claimed a tally of 13 cometary discoveries between 1760 and 1785, earning him the nickname "the comet ferret." During his routine sweeps across the skies, Messier sometimes chanced upon faint nebulae through his telescope eyepiece. Realizing that these dim patches of light in the realm of the stars far beyond the Solar System might confuse a comet hunter and lead to false announcements of cometary discoveries, Messier decided to compile a catalog containing a complete list of all known nebulae.

There were 45 objects featured in the first version of this deep-sky catalog; the final version published in 1781 had expanded to contain 103 objects, and seven more were added in the twentieth century following research uncovering further observations of Messier and his assistant Pierre Méchain (1744–1804). Méchain himself discovered 27 deep-sky objects, 18 of which made it into Messier's list.

Messier's list is still used today by astronomers eager to observe many of the brightest deep-sky objects visible from the northern hemisphere. It must be noted that the list is by no means exhaustive, and it neglects to include a number of reasonably bright and easily accessible deep-sky objects. John Dreyer's *New General Catalogue* of deep-sky objects, published in 1888, contains nearly 8,000 deep sky objects, hundreds of which are visible through a 15-cm telescope. In recent years the British amateur astronomer Sir Patrick Moore introduced the "Caldwell Objects," a list of similar scope to Messier's but covering deep-sky objects in the southern skies.

		The messier objects	
M	Constellation	Object type	Common name
1	Taurus	Supernova remnant	Crab nebula
2	Aquarius	Globular cluster	
3	Canes Venatici	Globular cluster	
4	Scorpius	Globular cluster	
5	Serpens	Globular cluster	
6	Scorpius	Open cluster	Butterfly cluster
7	Scorpius	Open cluster	Ptolemy's cluster
8	Sagittarius	Diffuse nebula	Lagoon nebula
9	Ophiuchus	Globular cluster	
10	Ophiuchus	Globular cluster	
11	Scutum	Open cluster	Wild Duck cluster

(continued)

(continued)

M	Constellation	The messier objects	
		Object type	Common name
12	Ophiuchus	Globular cluster	
13	Hercules	Globular cluster	Great cluster
14	Ophiuchus	Globular cluster	
15	Pegasus	Globular cluster	
16	Serpens	Nebula and open cluster	Eagle nebula
17	Sagittarius	Diffuse nebula	Omega nebula
18	Sagittarius	Open cluster	
19	Ophiuchus	Globular cluster	
20	Sagittarius	Diffuse nebula	Trifid nebula
21	Sagittarius	Open cluster	
22	Sagittarius	Globular cluster	Sagittarius cluster
23	Sagittarius	Open cluster	
24	Sagittarius	Open cluster	Sagittarius star cloud
25	Sagittarius	Open cluster	
26	Scutum	Open cluster	
27	Vulpecula	Planetary nebula	Dumbbell nebula
28	Sagittarius	Globular cluster	
29	Cygnus	Open cluster	
30	Capricornus	Globular cluster	
31	Andromeda	Spiral galaxy	Andromeda galaxy
32	Andromeda	Elliptical galaxy	Satellite of M31
33	Triangulum	Spiral galaxy	Triangulum galaxy
34	Perseus	Open cluster	
35	Gemini	Open cluster	
36	Auriga	Open cluster	
37	Auriga	Open cluster	
38	Auriga	Open cluster	
39	Cygnus	Open cluster	
40	Ursa Major	Double star	Winecke 4 (WNC 4)
41	Canis Major	Open cluster	
42	Orion	Diffuse nebula	Orion nebula
43	Orion	Diffuse nebula	De Mairan's nebula
44	Cancer	Open cluster	Praesepe
45	Taurus	Open cluster	Pleiades
46	Puppis	Open cluster	
47	Puppis	Open cluster	
48	Hydra	Open cluster	
49	Virgo	Elliptical galaxy	
50	Monoceros	Open cluster	

(continued)

(continued)

		The messier objects	
M	Constellation	Object type	Common name
51	Canes Venatici	Spiral galaxy	Whirlpool galaxy
52	Cassiopeia	Open cluster	
53	Coma Berenices	Globular cluster	
54	Sagittarius	Globular cluster	
55	Sagittarius	Globular cluster	
56	Lyra	Globular cluster	
57	Lyra	Planetary nebula	Ring nebula
58	Virgo	Spiral galaxy	
59	Virgo	Elliptical galaxy	
60	Virgo	Elliptical galaxy	
61	Virgo	Spiral galaxy	
62	Ophiuchus	Globular cluster	
63	Canes Venatici	Spiral galaxy	Sunflower galaxy
64	Coma Berenices	Spiral galaxy	Black Eye galaxy
65	Leo	Spiral galaxy	
66	Leo	Spiral galaxy	
67	Cancer	Open cluster	
68	Hydra	Globular cluster	
69	Sagittarius	Globular cluster	
70	Sagittarius	Globular cluster	
71	Sagitta	Globular cluster	
72	Aquarius	Globular cluster	
73	Aquarius	Asterism	
74	Pisces	Spiral galaxy	
75	Sagittarius	Globular cluster	
76	Perseus	Planetary nebula	Little Dumbbell nebula
77	Cetus	Spiral galaxy	
78	Orion	Diffuse nebula	
79	Lepus	Globular cluster	
80	Scorpius	Globular cluster	
81	Ursa Major	Spiral galaxy	Bode's galaxy
82	Ursa Major	Irregular galaxy	Cigar galaxy
83	Hydra	Spiral galaxy	Southern Pinwheel
84	Virgo	Spiral galaxy	
85	Coma Berenices	Spiral galaxy	
86	Virgo	Elliptical galaxy	
87	Virgo	Elliptical galaxy	
88	Coma Berenices	Spiral galaxy	
89	Virgo	Elliptical galaxy	

(continued)

(continued)

M	Constellation	Object type	Common name
		The messier objects	
90	Virgo	Spiral galaxy	
91	Coma Berenices	Spiral galaxy	
92	Hercules	Globular cluster	
93	Puppis	Open cluster	
94	Canes Venatici	Spiral galaxy	
95	Leo	Barred spiral galaxy	
96	Leo	Spiral galaxy	
97	Ursa Major	Planetary nebula	Owl nebula
98	Coma Berenices	Spiral galaxy	
99	Coma Berenices	Spiral galaxy	
100	Coma Berenices	Spiral galaxy	
101	Ursa Major	Spiral galaxy	Pinwheel galaxy
102	Draco	Spiral galaxy	
103	Cassiopeia	Open cluster	
104	Virgo	Spiral galaxy	Sombrero galaxy
105	Leo	Elliptical galaxy	
106	Ursa Major	Spiral galaxy	
107	Ophiuchus	Globular cluster	
108	Ursa Major	Spiral galaxy	
109	Ursa Major	Spiral galaxy	
110	Andromeda	Elliptical galaxy	Satellite of M31

Project #13: View the Top Ten Messier Objects

The ten deep-sky objects featured here have been chosen for their variety and challenge. Some are easily seen through binoculars, while others require dark skies and a telescope to discern.

M31

More than 2.5 million light years distant, the Andromeda Galaxy is the biggest member of the Local Group of galaxies. It is easy to locate and can be seen without difficulty with the unaided eye from dark suburban sites. We see M31 from an angle of around 30° above its plane, so that it is somewhat foreshortened. Through binoculars it appears as a bright misty oval around half degree wide, and from dark sites it stretches yet further across the field of view. A 20-cm telescope will reveal hints of structure within the galaxy, including a prominent dark lane and a suggestion of knottiness (large nebulae) in its spiral arms. Nearby are its small satellite galaxies, M32 and M110, both visible as tiny condensed blobs through 80-mm binoculars. M32, the brighter of the pair, lies around half a degree south of M31's center, while M110 lies around a degree to the northwest.

Telescopic observation of M31, 20-cm Schmidt-Cassegrain Telescope (Credit: Peter Grego)

M57

The Ring Nebula is one of the sky's best known planetary nebulae. It can easily be found since it lies almost directly between Beta and Gamma Lyrae. M57 is rather small in apparent diameter, binoculars showing it as an almost star-like point of light. Through a telescope at a medium to high magnification it resembles a sharply defined glowing smoke ring.

Observation of M57, 20-cm SCT (Credit: Peter Grego)

M1

Located just over one degree north of Zeta Tauri, the faintly
glowing supernova remnant of the Crab Nebula requires at least
an 8-cm telescope to be seen well. Even through large instruments
it appears as a gray, rather featureless elliptical patch. Once found,
it's thrilling to know that you're looking directly at the remnants
of a stellar explosion seen way back in 1054.

M42

The Orion Nebula is one of the biggest and brightest nebulae in
the heavens. Considerable structure within the nebula can be seen
with binoculars alone, and a small telescope will reveal a glowing
greenish mass with delicate wisps, intruded upon by a prominent
dark lane. Several stars can be seen in and around the nebula,
notably the Trapezium (Theta Orionis), a bright quadruple star.
Larger telescopes will show breathtaking detail within the nebula.
Just to its north lies the fainter De Mairan's Nebula (M43), with a
single bright star nestling at its center.

Telescopic observation of M1, 30-cm reflector (Credit: Peter Grego)

Telescopic observation of M42, 30-cm reflector (Credit: Peter Grego)

M51

Famous for its depiction by William Parsons using the "Leviathan of Parsonstown" (see below), the Whirlpool Galaxy is an easy binocular object. A 20-cm telescope is required to reveal a hint of the object's spiral arms. M51's companion galaxy, NGC 5195, is visible through a 10-cm telescope and is joined to M51 by a faint bridge that can be discerned through larger instruments.

Telescopic observation of M51, 20-cm SCT (Credit: Peter Grego)

M45

A handful of the brightest stars within the Pleiades star cluster are easy to see with the unaided eye. Binoculars will reveal many dozens of young blue stars, including a lovely arc of stars running south from Eta Tauri (Alcyone) and from dark skies a hint of reflecting nebulosity may be glimpsed near 23 Tauri (Merope) – traces of the interstellar cloud through which the cluster is currently passing.

Lunar occultation of the Pleiades, observed in August 1988, 6-cm refractor (Credit: Peter Grego)

M27

The Dumbbell Nebula is the finest planetary nebula in the northern skies. Easily visible through binoculars as a large, well-defined patch amid a rich star field, M27 resembles a broad misty dickie bow through the telescope; its greenish hue is noticeable through larger instruments.

M8

A superb diffuse nebula, the Lagoon Nebula is visible without optical aid as a Moon-sized glow in Sagittarius. Telescopes show a dark band within M8, dividing it in two, and a structure that John Herschel called the "Hourglass Nebula". NGC 6530, an open cluster, shines within the nebula's eastern glow.

M104

Located in the southwest of Virgo, the Sombrero Hat Galaxy is a bright eighth magnitude edge-on galaxy whose nuclear bulge rises

Observation of M27, 20-cm SCT (Credit: Peter Grego)

This 15×70 binocular observation is centered on M8 (Credit: Peter Grego)

Observation of M104, 30-cm reflector (Credit: Peter Grego)

smoothly on either side of its spindly spiral arms. A dark dust lane running through the center of M104 is visible through a 30-cm telescope.

M13

Hercules' Great Cluster is the northern sky's brightest globular cluster. Lying just 2.5° south of Eta Herculis, M13 is easy to locate. Indeed, it is faintly visible with the unaided eye. Binoculars show it to be an extensive fuzzy patch around half the apparent diameter of the full Moon. Viewed through a 15-cm or larger telescope, the cluster is an amazing sight; the brightest of its 300,000 outlying stars can be resolved, and these appear to be arranged in several distinct radial lines. Hints of darker lanes can be discerned within the cluster's outer regions; photographs don't show these features well, but our perception through the eyepiece produces a different impression.

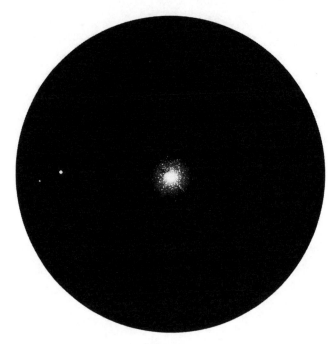

Observation of M13, 20-cm SCT (Credit: Peter Grego)

The Herschel Sky Surveys

In his famous reviews of the heavens commencing in 1783, William Herschel swept the northern skies in a systematic manner using a 47.5-cm reflector of his own construction; with the help of his sister Caroline he charted hundreds of previously unknown star clusters and nebulae. His careful observations enabled him to describe the general form of our Galaxy and even led to the discovery of the solar apex near Lambda Herculis – the point in space towards which the Sun, carrying with it the Solar System, was heading. Completed in the space of just a year and a half, Herschel's first survey included a catalog of around a thousand previously unknown deep-sky objects. Further surveys were conducted in 1789 and 1802, bringing his total of deep-sky discoveries to a staggering 2,500 objects.

John Herschel carried on his father's deep-space survey. Having made additional surveys of the northern skies, in 1834 he traveled with his family to Feldhausen at the foot of Table Mountain in South Africa to commence a survey of the southern skies using the same cumbersome "20 foot" telescope with which his father had conducted his own sky survey, in addition to a more convenient equatorially mounted 125-mm refractor. After his return to England and 7 long years of reducing his observational data, those 4 years of hard work in South Africa led to the publication of the *Cape Results* in 1847, the apex of John Herschel's astronomical career.

Aberration of Light

In 1728, James Bradley (1693–1762) discovered that the observed position of the stars on the celestial sphere appeared to vary very slightly with the motion of the Earth around the Sun. Every star appeared to move in a tiny self-centered ellipse each year, the flatness of the ellipse dependent on the star's latitude. This phenomenon, known as the aberration of light, occurs because

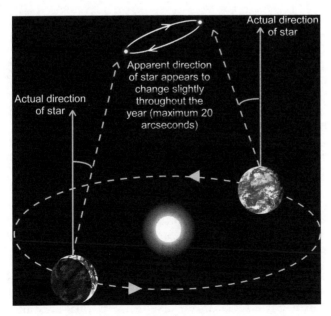

This shows the aberration of starlight caused by Earth's rotation around the Sun (Credit: Peter Grego)

Earth, traveling at a speed of 30 km/s in its orbit around the Sun, intercepts starlight traveling at 300,000 km/s. Imagine running through a shower of rain; even though the rain is actually falling straight down, you need to tilt your umbrella further forward to stop the rain from hitting your head, since the raindrops appear to be approaching from a slanted direction when running.

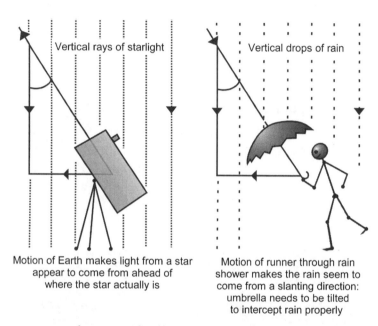

Motion of Earth makes light from a star appear to come from ahead of where the star actually is

Motion of runner through rain shower makes the rain seem to come from a slanting direction: umbrella needs to be tilted to intercept rain properly

In the same way that an umbrella needs to be tilted at an angle when running through the rain – the drops appearing to come from a slanting angle when in fact they are falling vertically – the direction in which a telescope points to center a star is slightly different from a star's actual direction, by virtue of the aberration of starlight caused by the Earth's motion through space (Credit: Peter Grego)

Plumbing Cosmic Depths

Following up on Bradley's work in charting the position of stars, Friedrich Bessel (1784–1846) was the first to observe the regular change of a star's position due to the changing line of sight as Earth orbits the Sun, an effect known as stellar parallax. Stars nearer to Earth will show the greatest parallax, while the parallax of increasingly distant stars is more difficult to measure. Bessel's measurement of the parallax of the star 61 Cygni in 1838 allowed its distance to be calculated – the first measurement of the distance to the stars. It was found that 61 Cygni was an incredible 11 light years away.

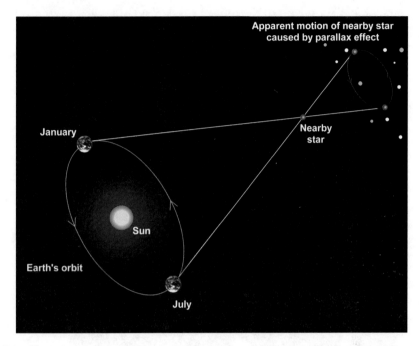

This illustrates the apparent motion of a nearby star against the distant celestial background, caused by our changing line of sight. The effect is greatly exaggerated for clarity

A Small Galaxy in a Big Universe

Comfortable notions that the entire Universe was dominated by the Milky Way began to be challenged during the late nineteenth century. Photography began to reveal more of deep space than could be seen by the visual observer, and the new science of spectroscopy laid bare the composition of stars and nebulae.

Even the keenest visual observer using a large telescope will struggle to discern detail within most of the nebulae that dot the heavens. Using the "Leviathan" at Birr Castle in Ireland – a telescope with a huge 1.8-m diameter mirror that was then the largest telescope in the world – William Parsons visually observed nebulae, hoping to discern individual stars within them. In 1845 he discovered the spiral nature of one particularly bright nebula, now known as the Whirlpool Galaxy. Other nebulae were noted to have a spiral form, too. Some of these, like the Whirlpool, appeared face-on, while others were presented at a variety of angles, including some that were almost edgewise.

The Whirlpool Galaxy, sketched in 1845 by William Parsons at the eye-piece of the 1.8 m "Leviathan" at Birr Castle, Ireland

When photography began to be used extensively for deep sky studies from the late nineteenth century onwards, the true grandeur of many deep sky objects was uncovered for the first time. Deep sky objects were divided into a number of distinct groups – open star clusters, globular star clusters, amorphous gaseous nebulae, planetary nebulae, and spiral nebulae. Astronomers began to question whether the nebulae really did represent stages in the evolution of distant solar systems – especially since spiral nebulae had been shown to be very large objects packed with faint stars and at a great distance from Earth.

In 1908 Henrietta Leavitt (1868–1921) carefully studied photographs of the Milky Way's near neighbor, the Small Magellanic Cloud. She identified a number of stars whose luminosity varied

Chromolithograph showing the "Leviathan of Parsonstown" in its prime – William Parsons' 1.8 m reflector at Birr Castle, Ireland (Credit: Peter Grego)

over a regular cycle; the period of time over which they varied from minimum to maximum brightness appeared to depend on their actual luminosity. Leavitt was able to state this with reasonable certainty, since it could be assumed that the stars within the Small Magellanic Cloud are all at roughly the same distance from us.

Soon, these so-called Cepheids in our own Milky Way were identified; this was of immense importance, since by knowing the actual brightness of a star its true distance could be measured. Now astronomers had a tool with which to measure the vast distances between stars, star clusters, and galaxies. After studying Cepheids within the globular star clusters that surround the Milky Way, Harlow Shapley (1885–1972) produced the first map of our Galaxy. Shapley's Milky Way was a vast system of a hundred billion stars, arranged in a flattened disk; the Solar System appeared to be located some distance from the Galactic center.

Observations of Cepheid variable stars allow astronomers to gauge the scale of the Milky Way and the distance of globular clusters, the Magellanic Clouds, and more distant galaxies (Credit: Peter Grego)

Hubble's Universe

In the 1920s Edwin Hubble (1889–1953) used the 2.5-m telescope at Mount Wilson in California to search for Cepheid variable stars within a number of spiral nebulae. Questions about their size and distance had been asked for decades, and Hubble finally provided measurements that astounded astronomers. Far from being residents of our own Milky Way, objects such as the spiral nebulae in Andromeda and Triangulum were shown to be independent galaxies of comparable size to the Milky Way, each made up of thousands of millions of stars and located at least 25 times further than the most distant stars of our own Galaxy. It was a sobering thought that the light from these galaxies had been emitted in prehistoric times, millions of years ago, long before the first modern humans had evolved.

By studying the light from distant galaxies, Hubble went on to conclude that the Universe is expanding, and that the further away a galaxy is, the faster it appears to be moving. To understand how Hubble arrived at this remarkable conclusion, let's take a look at a useful tool of astronomers, the spectroscope. Spectroscopes use a prism to split light into its component colors. Narrow dark bands called Fraunhofer lines can be seen within the resulting rainbow of light. These are produced by certain chemical elements within the stars. The Fraunhofer lines in the spectrum of an object moving away from us at a very high velocity will appear to be shifted towards the red end of the spectrum; this is called redshift. Hubble showed that the further a galaxy is from us, the greater its redshift; in fact, galactic redshifts increase in proportion to their distance from us. It was later shown that galaxies are distributed throughout space in clusters and superclusters. Our own Galaxy was shown to be just one of billions of galaxies spread throughout space and time.

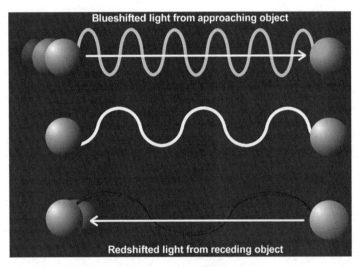

Wavelengths of light from approaching objects are compressed, shifted towards the blue end of the spectrum, while light from objects receding from the observer is stretched out into the *red* (Credit: Peter Grego)

The Big Bang and the Expanding Universe

Of course, the observation that galaxies are all moving away from us is simply caused by our cosmic perspective within an expanding Universe; the same observation could actually be made from any point within the Universe. Imagine a semi-inflated balloon upon which are drawn a number of dots; these dots represent galaxies. As the balloon (representing the fabric of space-time) is inflated further, and its surface expands, the dots all move away from each other. There's no central point of expansion on the spherical surface of the balloon. In the same way, there's no central point in this expanding Universe of ours, away from which everything is moving.

If the Universe is expanding, then it follows that there must have been a time in the remote past when everything in the Universe was contained within a single point. This point – the primordial atom – exploded to create time, space, and the entire Universe. Cosmologists are now fairly confident that this event, known as the Big Bang, took place between 13 and 14 billion years ago.

One of the best pieces of evidence supporting the Big Bang theory comes in the form of an echo from the beginning of time – the cosmic microwave background radiation (CMB), discovered in 1965 by U. S. radio astronomers Arno Penzias and Robert Wilson, a discovery for which they won the Nobel Prize. Their horn-shaped radio antenna at the Bell Laboratories in New Jersey picked up

The Bell Laboratories horn antenna with which the "echo" of the Big Bang was discovered in the form of the cosmic microwave background (CMB) radiation

a diffuse background signal that appeared to emanate from all directions uniformly. Puzzled by the signal, Penzias and Wilson eliminated all possible mechanical and instrumental causes – including pigeon droppings in the antenna – yet the signal stubbornly remained. Eventually the astronomers concluded that the radio signal was nothing less than a relic of the hot primordial fireball, dating to around 380,000 years after the Big Bang. Originally a temperature of 3,000 K, the expansion of the Universe has redshifted the radiation into microwave wavelengths, where it permeates the Universe at a temperature of just 3 K (3°C above absolute zero).

Ripples in the Cosmic Fabric

Just 25 years after the discovery of the CMB, the Cosmic Background Explorer satellite detected minute fluctuations in its intensity. Confirmed by other more sensitive instruments, these ripples within the CMB are, in effect, the first signs of the large-scale changes in the once-smooth fabric of space that led to the formation of galaxies.

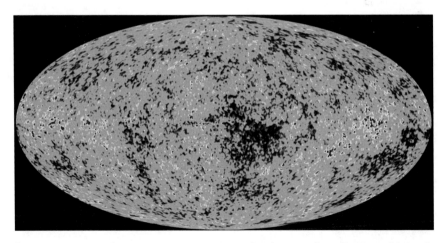

Fluctuations in the cosmic microwave background radiation show the seeds of galaxy formation in the early Universe. (Photo courtesy of NASA)

6. Beyond Vision

As we have seen, from the time of Galileo right through to the early nineteenth century, astronomers were restricted to viewing the Universe in visible light by peering through the telescope eyepiece; sketching, careful measurement, and written notes were the only practical means by which telescopic observations were capable of being made. This long-established *status quo* was first challenged with the invention of a certain "magic box."

Photographic Memories

Since ancient times it has been known that a chink of light entering a darkened room produces an upside-down image of a bright external scene on the wall opposite the opening; the smaller the aperture, the dimmer but sharper the projected image becomes. This phenomenon is even noticeable in broad daylight, when multiple images of an eclipsed Sun are haphazardly projected onto the ground through gaps between the branches of a tree. Greek philosopher Aristotle (384–322 BCE) noticed this effect by chance, experiencing a "tree-based revelation" in much the same way that Newton is said to have hit on the idea of gravity after seeing an apple fall to the ground. Aristotle proceeded to construct his own camera obscura (Latin for "darkened room") with a single small hole to allow sunlight to enter. He noted that the Sun's projected image always appeared circular, regardless of the small hole's shape. The projected image was therefore a real representation of the Sun's disk.

Iraqi-born Abu Ali Al-Hasan Ibn al-Haitham (known in the West as Alhazen, 965–1039 CE), performed many practical experiments on optics using sound scientific methods, and wrote about them in his remarkable *Kitâb al-Manâzir* (*Book of Optics*, 1011–1021 CE). Alhazen – who must have come extremely close to discovering the principles of the telescope some 600 years before

P. Grego and D. Mannion, *Galileo and 400 Years of Telescopic Astronomy*, 207
Astronomers' Universe, DOI 10.1007/978-1-4419-5592-0_6,
© Springer Science+Business Media, LLC 2010

Lippershey – describes the camera obscura and the principles involved in image projection.

Project #14: Construct a Pinhole Camera Obscura

A simple camera obscura is quick and easy to make out of everyday household objects; no specialist tools are required, and it doesn't take any great skill to produce. We're looking at making a box with a small hole at one end and a translucent screen covering the other end. The box itself can be small, such as an empty tin can or a potato chips tube, or it can be larger, like a shoebox or an empty paint can.

Make sure the inside of your box or tube is clean and dry, and if its interior is in any way bright or reflective, paint the inside black to minimize internal reflections. If you're using a tin can, remove both ends safely; if using a cardboard box, remove one side completely. On the side opposite the open side, create a fingernail-sized hole in the middle. This hole doesn't have to be perfect, as your next step is to completely cover it with a sheet of tin foil, firmly attaching it with clear tape. Next, using a needle prick a small hole in the centre of the tin foil. Cover the open side of the box or tube with a sheet of tissue paper or tracing paper, stretching it so that it lies taut and flat, and secure it in place with tape or an elastic rubber band.

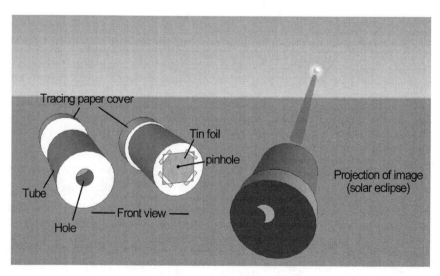

Construction of a simple camera obscura (Credit: Peter Grego)

The image projected onto the screen is viewed from the outside, where it appears upside-down. Brightly lit external landscape scenes can be viewed with more clarity if your head is covered with a coat and the camera obscura is pointing out of it.

Simple devices like this are useful for astronomy; solar eclipses can be viewed safely with them, and it's also possible to discern some of the larger sunspots and the rare transits of Venus with one. Try it at night with the full Moon. Can you discern the dark patches of the lunar seas and one or two of the brighter spots surrounding the larger impact craters?

It wasn't until the eighteenth century that the eye-opening effects produced by the camera obscura really took off, when mirrors were used to divert the incoming light to produce a correctly orientated image of an external scene in a darkened chamber. Such devices were not only used for entertainment; artists also used them to reproduce a scene as precisely as possible by lightly sketching over the projected image and using this as a basis for a painting. Examples of such public camera obscuras can be found in many places around the world; for example, there is one at the old Royal Observatory in Greenwich that projects a marvelous view of the city of London onto a large table.

Going one small, logical, and historically important step further from the camera obscura, in 1822 the physicist Joseph Niépce (1765–1833) used a camera obscura to project an image upon a pewter plate covered with a photosensitive material called bitumen of Judea, a chemical which hardens when exposed to light. Developing the plate in a dark room, the still soft chemical was washed away, producing a permanent negative image. When the plate was carefully coated with ink (so that only the raised areas are inked) and pressed against a sheet of paper, a positive print was created.

It might be said that the development of photography marked the beginning of true human extrasensory perception. The word itself, meaning "writing with light," derives from the ancient Greek for light ("phos") and writing ("grapho") and was coined in 1834 by Hercules Florence (1804–1879). In these days of instant digital photography we still immediately associate with photography the sight of a person leaning over a tripod, the photographer's upper body covered by a black cloth to eliminate stray light and make the projected image clearer to see. Indeed, photographers using large-format view cameras still make use of the principle of the camera obscura to project and focus a scene onto a ground glass (which appears inverted and flipped to the photographer), and covering cloths are still used to this day.

Milestones in Astrophotography

The long-established dominance of purely visual work at the telescope eyepiece met its first challenge in 1839 when John Draper (1811–1882) used a 12-in. telescope to secure the world's first astronomical photograph – an image of the Moon that required no less than 20 min of exposure time. Draper's images were known as Daguerrotypes, an early form of permanent photography based on a process developed by Niépce's business partner Louis Daguerre (1787–1851). Although rather grainy and lacking in fine detail – equivalent to the sort of image that can be taken today by pointing a disposable camera through the eyepiece of a small telescope – these pioneering astrophotographs demonstrated that the human eye now had a powerful new rival with enormous potential to change telescopic astronomy.

Daguerrotypes are one-off images; small and difficult to view because of their mirror-like properties, they can't be reproduced

An early lunar image, from a Daguerrotype taken by English-born American chemist John Draper through a 12-in. telescope

directly. Following pioneering research by legendary astronomer John Herschel (who coined the terms negative and positive as applied to photographic images) and English scientist William Fox Talbot (1800–1877), who was instrumental in creating a form of photography that could be mass produced, an unlimited number of exact copies of photographic images on paper could be made. In 1839 Fox Talbot invented the Calotype, announcing it in his paper to the Royal Society as "the art of photogenic drawing, the process by which natural objects may be made to delineate themselves without the aid of the artist's pencil." "How charming it would be," he wrote, "if it were possible to cause these natural images to imprint themselves durably and remain fixed on the paper!"

Although initially inferior in quality to the Daguerreotype, the potential of producing paper prints from glass-plate negatives blossomed in the ensuing years, and in 1844 Fox Talbot published a photographically illustrated book charmingly titled *The Pencil of Nature*. Fox Talbot's work led directly to the great advancements in photography during the nineteenth and twentieth century, while the Daguerrotype was to remain a specialist form of bespoke art (it still has its practitioners to this day).

Promising to achieve a level of positional accuracy unable to be matched by a human straining at the eyepiece with a micrometer, photography was able to image large areas in a relatively short space of time, and being a cumulative process, long exposures were capable of revealing objects too faint to be detected with the eye. Although sketchers of astronomical objects might strive to be as accurate as possible in their rendition of, say, a lunar crater, a degree of artistic license – consciously or subconsciously applied – is always bound to a certain extent to degrade an observational drawing's accuracy and its scientific value. Not so with the photographic image, and although it's not entirely true to claim that the camera never lies, it is certainly a more trustworthy witness.

Astronomer Edward Emerson ("Eagle-Eyed") Barnard (1857–1923) put the case for the effectiveness of photography after seeing the remarkable variation between different observers' drawings of the Sun's corona seen at the total solar eclipse of July 26, 1878, and published by the U. S. Naval Observatory. "On examination scarcely any two of them would be supposed to represent the same object, and none of them closely resembled the photographs

The first correctly exposed photograph (a Daguerrotype) of a solar eclipse –
that of July 28, 1851, taken by Mikhail Berkowski at the Royal Observatory
in Königsberg, Prussia, through a 60-mm refracting telescope attached to
the 158-mm Fraunhofer heliometer. The exposure was 84 s and shows the
solar corona at totality

taken at the same time," he wrote. "The method of registering
the corona by free-hand drawing under the conditions attending a
total eclipse received its death-blow at that time, for it showed the
utter inability of the average astronomer to sketch or draw under
such circumstances what he really saw."

Numerous well-equipped amateur astronomers at the cutting-
edge of nineteenth-century astronomy were among the most noteworthy practitioners of astrophotography. As we shall see, a number
of highly skilled amateurs continue to produce images that often
match those taken at professional observatories.

Following in the pioneering astrophotographic footsteps of his
father, Henry Draper (1837–1882) secured the world's first deep-
sky image on September 30, 1880. Showing the majestic Orion
Nebula, the photograph was taken through his 275-mm (11-in.)
Clark refractor at his observatory in Hastings-on-Hudson in New
York and required an exposure time of 50 min. Two years later he
secured a more detailed photograph of the Orion Nebula with an
exposure time of more than 2 h; the dimmer nearby nebula M43 is
also clearly shown on this image. Henry Draper also took around
1,500 images of the Moon, some of which remained unsurpassed

for many years. Of course, the Moon was, and remains, a popular target for the astrophotographer because of its sheer brightness (permitting relatively short exposures with small telescopes) and the high level of detail visible.

Henry Draper's astronomical observatory at Hastings-on-Hudson in New York

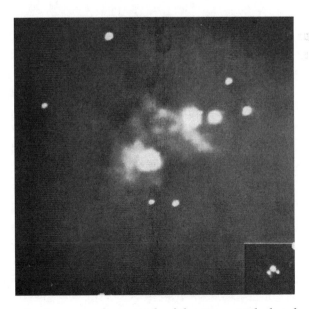

The first deep-sky image, a photograph of the Orion Nebula taken by Henry Draper in 1880

One of Henry Draper's many photographs of the Moon, this one taken on September 3, 1863

After abandoning his New York lawyer's practice, Lewis Rutherfurd (1816–1892) took due diligence in pursuing his love of amateur astronomy, proving himself a *force majeure* in pioneering astrophotography. Faced with a dearth of off-the-shelf equipment Rutherfurd devised much of his own astronomical gear, inventing a micrometer for astrometric work on photographs and improved on a ruling machine to produce diffraction gratings (giving 17,000 lines per inch) for making detailed observations of stellar spectra.

Dissatisfied with using regular telescopes for astrophotography, Rutherfurd teamed up with Henry Fitz (1808–1863) to design and build the first telescopes intended specifically for astrophotography; since photographic plates were most sensitive to blue light, these photovisual instruments required their objective lenses to be corrected accordingly. Before long, photovisual instruments were installed at virtually every major observatory in the world. Between 1865 and 1875 Rutherfurd secured photographs of the Moon, many of which remained unsurpassed until the beginning of the twentieth century. He also produced many high-quality photographs of the Sun and planets, as well as deep-sky objects.

Notable astrophotographic firsts were achieved by Pierre Janssen in 1881, with the first detailed image of a comet (which later became known as the Great Comet of 1882), and in 1885 by brothers Paul and Prosper Henry, who took the first successful images of Jupiter and Saturn.

The 28-in. photovisual refractor at the old Royal Observatory, Greenwich, completed in 1893, whose giant lens was to allow the telescope to perform as an observational and photographic instrument (Credit: Peter Grego)

In the UK several giants of Victorian amateur astronomy made significant contributions to astrophotography. Isaac Roberts (1829–1904), a wealthy amateur astronomer, used a 508-mm (20-in.) Grubb photographic reflector to take breathtaking images of celestial objects. Many of his long exposure deep-sky photographs

One of Lewis Rutherfurd's spectacular lunar photographs, this one taken in the 1870s and showing a staggeringly favorable libration for the Moon's eastern limb

The Great Comet of 1882, photographed by David Gill from South Africa

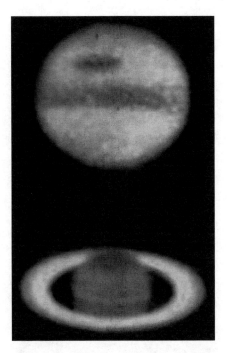

Jupiter and Saturn imaged in 1885 by the Henry brothers. Note the intensity of Jupiter's Great *Red* Spot and the Equatorial Zone

were published in his books *A Selection of Photographs of Stars, Star-Clusters and Nebulae* (Vol. 1, 1890, Vol. 2, 1899). Roberts' photograph of the Andromeda Galaxy (then known as the Great Nebula in Andromeda) surprised and delighted the astronomical world by clearly revealing that that object was a spiral nebula. The spiral form of some nebulae, notably M51 (the Whirlpool Galaxy) in Ursa Major, had first been noted visually in 1845 by William Parsons using his 1.83-m (72-in.) "Leviathan" reflector at Birr Castle in Ireland.

So good were Roberts' deep-sky images that professional astronomers increasingly began to look towards reflectors as being the way forward if the promise held out by astrophotography was to be realized in full. With their bigger apertures, reflecting telescopes were more suited to long-exposure astrophotography; the bigger the telescope, the greater its light-gathering power – and the photographic plate was exceedingly hungry for photons.

Reflecting telescopes had been somewhat neglected by professional astronomers during the first half of the nineteenth century, and the tendency was to plump instead for achromatic refractors

Isaac Roberts' groundbreaking photograph of the Andromeda Galaxy of 1888, which clearly shows its spiral nature

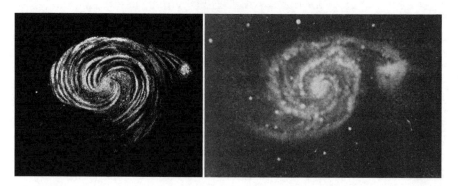

William Parsons' sketch of M51 compared with Isaac Roberts' image of the same object

with which to conduct their astronomical research. Following the invention of the achromatic objective lens by John Dolland in the mid-eighteenth century and Pierre Guinand's invention of bubble-free optical glass later that century, refractors came into their own, reaching their zenith of popularity in the mid-nineteenth century.

Refractors were considered better suited to making precise astrometric observations; they have no obstruction in their light path, and all things being equal they deliver views with more resolution and contrast. In terms of performance, a good refractor is equivalent to a reflector of larger size, so they require a smaller and more manageable mount. Moreover, refractors require less maintenance than reflectors; reflectors at that time used mirrors made out of polished speculum metal (an alloy of copper and tin), which needed to be frequently removed and repolished due to rapid tarnishing.

In 1856 Léon Foucault (1819–1868) invented a new form of telescope mirror – one which used a figured glass disk coated with a layer of silver. Cheaper and more effective than the specula of old, glass mirrors were capable of being made to sizes unattainable by objective lenses. In terms of cost and ease of manufacture, the lens is trumped by the mirror. An achromatic lens requires high quality optical glass and four precisely figured surfaces, while a mirror need not be made from optical glass and has only one curved surface with which to focus the light.

Big reflectors were one of the astronomical loves of Andrew Ainslie Common (1841–1903). At his home beneath the (then) tolerably dark skies of Ealing, London, he installed a 36-in. (910-mm) reflecting telescope, containing one of the first really big silvered glass mirrors. Common used this instrument to take an impressive photograph of the Orion Nebula in 1883, which won him the Gold Medal of the Royal Astronomical Society. After seeing this image, Agnes Clerke wrote: "Photography may thereby be said to have definitely assumed the office of historiographer to the nebulae; since this one impression embodies a mass of facts hardly to be compassed by months of labour with the pencil."

Those "many months of labour" alluded to by Clerke had certainly been put in by George Bond (1825–1865), ironically a pioneering astrophotographer, in producing his famous sketch of the Orion Nebula after dozens of careful visual observations made at the eyepiece of a 15-in. telescope.

Clearing the decks for an even larger telescope, Common soon sold the 36-in. mirror; it now serves in the Crossley reflector at the Lick Observatory in California, where, after a distinguished astrophotographic career, it is currently engaged in the hunt for extrasolar planets. Common then went on to construct a mammoth

1.52-m (60-in.) telescope that held the title of the world's biggest telescope between 1889 and 1908.

For most kinds of visual observation it's fair to say that a telescope drive needn't track the motion of the sky with unerring pinpoint precision to keep an object centered in the field for many hours on end. Edward Barnard, a great visual observer as well as a pioneering astrophotographer, realized that the alignment of a telescope and the accuracy of its drive needed to be absolutely spot-on in order to produce good astrophotographs requiring lengthy exposures. The rise of astrophotography (Barnard being one of its most masterful late nineteenth-century exponents and

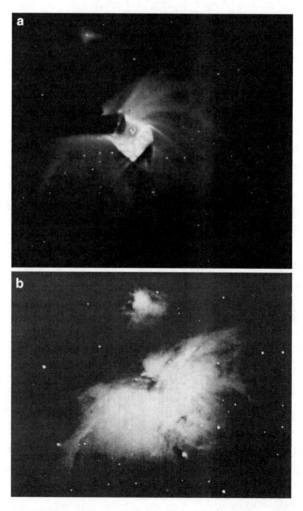

An observational sketch of the Orion Nebula by George Bond made in 1863, compared with the same object photographed in 1883 by Andrew Common

The Horsehead Nebula in Orion, photographed by Barnard

practitioners) led directly to improvements in telescope mounts and the design and accuracy of their drives. While compiling the first photographic atlas of the Galaxy (*A Photographic Atlas of Selected Regions of the Milky Way*, published in 1927) Barnard identified a host of hitherto unknown dark nebulae – opaque or translucent masses of interstellar dust and gas that block the background light – and went on to catalog no fewer than 366 of them. Perhaps the most famous of these dark nebulae is Barnard 33, the Horsehead Nebula in Orion.

Before the nineteenth century was out, every professional observatory on Earth was engaged in astrophotography. Since astrophotography mushroomed way beyond its humble beginnings, it's impossible to continue listing the achievements of ensuing generations of pioneering astrophotogaphers. A noteworthy milestone in astrophotography came with the invention of the Schmidt camera by Bernard Schmidt in 1930, a reflector capable of producing excellent wide-field images. Using the 1.2-m Oschin Schmidt camera at Mount Palomar Observatory, a complete photographic survey of the northern skies was published in the form of the *Palomar Sky Survey* (1954). Using instruments of the same focal length as the Oschin Schmidt – hence delivering images of identical field coverage – the Southern Sky Survey (1974–1990), the 1.2-m UK Schmidt Telescope at Siding Spring, and the European Southern Observatory's 1-m Schmidt camera completed our first coherent celestial photographic survey. Both surveys remain valuable resources to this day, and digitized versions are referred to by both amateur and professional astronomers.

The photographic plate remained firmly attached to the small end of professional telescopes until the 1980s, when CCD imagery began to supersede traditional photography. What the CCD gains in sensitivity – the CCD chip is several orders of magnitude more sensitive than the fastest film – it loses out in terms of field of view when compared to traditional photography.

Spectroscopy Shows Its True Colors

Going a significant step further than Newton, in 1800 William Herschel conducted a remarkable experiment in which sunlight was split into its constituent colors with a prism and a thermometer was dipped into each color to measure the temperature at various points in the spectrum. Quite unexpectedly he noted a sudden increase in energy lying in an unilluminated area just beyond the red end of the spectrum. Clearly an invisible form of energy mixed in with sunlight was producing this effect. Thus, infrared radiation was discovered, and the science of spectroscopy was born.

In 1814 Joseph von Fraunhofer (1787–1826) constructed the first accurate spectroscope; light entering the device through a narrow slit was collimated by an internal lens to produce parallel rays that were then split into a spectrum by a prism; the spectrum was observed through an eyepiece and could be measured against a reticle in the field of view. Careful examination of the spectrum produced in this manner revealed the presence of hundreds of narrow dark lines. As well as appearing in the solar spectrum, it was soon discovered that planets and stars displayed spectral lines, although stellar spectra were not all identical. Fraunhofer went on to map 574 lines in the Sun's spectrum, designating the main ones with the letters A to K and using other letters for less distinct lines. Now known as the Fraunhofer lines, it wasn't until 1859 that Gustav Kirchhoff (1824–1887) and Robert Bunsen (1811–1899) discovered their true significance.

In 1835 the French philosopher August Comte mused that there were limits to human knowledge, these limits being determined by the constraints on how much information it was possible to gather about any particular subject. Comte declared that the chemical composition of the stars was one such example of

Fraunhofer demonstrates the spectroscope

unobtainable knowledge. The stars were so unimaginably distant that it would never be possible to directly examine samples of star-stuff. Comte wrote: "On the subject of stars, all investigations which are not ultimately reducible to simple visual observations are...necessarily denied to us... We shall never be able by any means to study their chemical composition."

Comte's confident declaration was famously proven wrong just a few years later. After experimenting with chemical spectra produced in the laboratory Kirchhoff and Bunsen determined that each element has its own unique spectral signature, often likened to a "fingerprint." Hot luminous gas under low pressure produces bright emission lines at particular wavelengths, while the Fraunhofer lines in the solar spectrum are produced when light at certain wavelengths is absorbed by elements in the Sun's upper atmospheric layers. The wavelengths of the absorption lines are identical to the wavelengths of emission lines from the same gas when heated. Each of the 92 fundamental elements produces its own unique set of spectral lines,

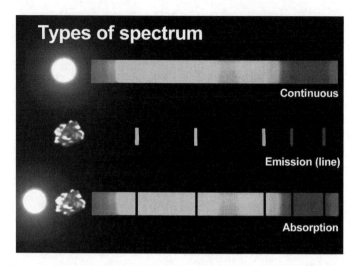

Types of spectrums: High-pressure gases such as a star emit a continuous spectrum (*top*); a low-density hot gas (*middle*) such as an emission nebula emits a line spectrum, each line or line set corresponding to particular chemical elements. An absorption spectrum (*bottom*) is produced when light with a continuous spectrum passes through a low-density cool gas and element-specific colors of light are absorbed, leaving dark lines in the spectrum (Credit: Peter Grego)

and their position relative to one another remains constant, whether the spectroscope is turned towards the Sun, a star thousands of light years away, or a galaxy billions of light years distant.

While making spectroscopic observations during the total solar eclipse of August 1868, Norman Lockyer (1836–1920) observed previously unknown lines in the yellow part of the Sun's spectrum – lines that were unable to be reconciled with any known element, as they had never been seen in laboratory experiments. Lockyer suggested that the yellow line was caused by an unknown element and named it "helium" after the Greek word for the Sun, *Helios.* Lockyer's new element was confirmed 25 years later when terrestrial helium was discovered.

More accurate spectrographs were developed when photography was applied to spectroscopy, with notable pioneering work at Harvard College Observatory by Edward Pickering (1846–1919), where great leaps forward were made in understanding stellar spectra. Pickering recruited many women to work for him, including Annie Jump Cannon (1863–1941), who developed a system of classifying stellar spectra. This has become the well-known series:

WOBAFGKMLT, in order of decreasing surface temperature. You might see in some older books spectral types R, N, and S after type M. These stars are giant carbon stars whose spectra show excess carbon and carbon compounds. Each letter is further sub-divided into ten divisions, e.g., B0 down to B9. Our own Sun is a G2 dwarf star or G2 V star, where Roman numerals I, II, III, IV, and V give rise to luminosity classes. Supergiants are I, bright giants II, normal giants III, sub-giants IV, and dwarfs or main sequence stars are V.

The harvard classification of stars

Surface temperature (Tsurface/K)	Peak wavelength (λ_{max}/nm)	Color of star	Harvard spectral classification
2,000	1450	Red	M
3,500	828	Orange	M
5,000	580	Yellow	K
6,000	483	Yellow	G
7,500	386	Yellow/white	F
10,500	276	White	A
11,500	252	Blue/white	B
20,000	145	Blue/white	B
40,000	72	Blue	O
80,000	36	Blue	W

There's a delightful mnemonic that can help you remember the sequence, which goes "Wow, Oh Be A Fine Girl/Guy, Kiss Me." However, the last three letters R N S of the original sequence have been superseded by C for carbon stars, so we no longer have the additional words "Right Now, Sweetie!" Classes L and T have been added recently in 1999 and denote the coolest stars with surface temperatures L (1,300–2,000 K) and T (below 1,200 K), but so far no one has come up with a mnemonic encompassing these two letters, so there's room for some creativity here.

The higher a star's surface temperature the shorter the peak wavelength, and above 8,500 K the asymmetric peak is in the ultraviolet. This ensures that stars with surface temperatures above 30,000 K look very blue, for example Alnitak (Zeta Orionis), which is an O9 blue supergiant, or even Rigel (Beta Orionis), a type B8 blue–white supergiant with a temperature of ~11,000 K.

Stars vary quite considerably in size, so a star on the main sequence compared to a red giant or red supergiant, all with the same surface temperature, could have radii in the ratios 1:50:500, so luminosities which scale as radius squared will be 1:2,500:250,000.

Below is a table giving rough values of radii, mass, luminosity, and main sequence lifetimes of stars in terms of the Sun's values R⊙, M⊙, L⊙.

Characteristics of stars on the Hertzsprung Russell diagram

	Main sequence	Giant	Supergiant
Radii R/R.	0.1–20	7–100	30–1,000
Mass M/M.	0.06–50	3–8	>8 up to 100
Luminosity L/L.	0.00001–200,000	6–2,000	8,000–1 million
Lifetimes in Myrs	0.5–1,000,000	2–1,000	2–20

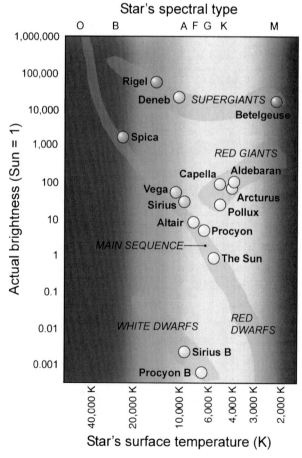

The Hertzsprung-Russell diagram (Credit: Peter Grego)

The electromagnetic spectrum
(Credit: Peter Grego)

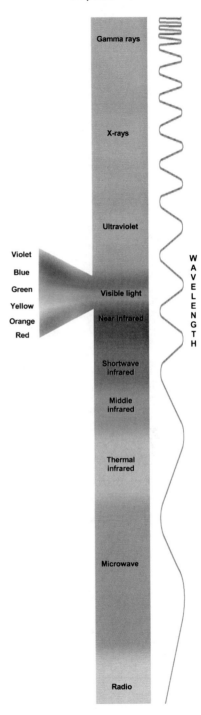

Fantastic Realms Beyond Vision

Twenty-first century astronomers are capable of exploring fantastic realms that tell us so much more about the Universe than is revealed in visible light alone. We now know that visible light is just a very tiny part of the electromagnetic (EM) spectrum, which consists of seven named regions spanning a vast range of wavelengths, from radio waves, which can have wavelengths hundreds of kilometers long, to ultra-high energy gamma rays with wavelengths smaller than a proton (less than 10^{-15} m long). Moving in order of wavelength, from longest to shortest, the EM spectrum consists of: radio, microwave, infrared, visible light, ultraviolet, X-ray, and gamma ray wavelengths. Fortunately, astronomers of the twenty-first century make good use of all of these wavelengths to explore the known Universe.

First proposed by James Clerk Maxwell (1831–1879) when he unified the electric and magnetic forces in 1869, EM waves consist of oscillating electric and magnetic fields that can radiate energy in the vacuum of space. Twenty years later, Heinrich Hertz (1857–1894) demonstrated radio waves, and in 1912 Guglielmo Marconi (1874–1937) was the first to send and receive radio transmissions across the Atlantic Ocean. The first radio telescopes were pioneered in the United States during the 1930s by Karl Jansky (1905–1950) and Grote Reber (1911–2002). Apart from visible light and radio waves, most other EM wavelengths can't penetrate Earth's atmosphere, so astronomers had to wait until the development of balloon-based, rocket-borne, and satellite observatories before the EM spectrum was further opened up to scrutiny.

Electromagnetic Revelations

Let's first look at the characteristics of electromagnetic waves and then see what each region of the EM spectrum offers the astronomer in observing the Universe. EM waves all travel at the same velocity in space and consist of oscillating electric and magnetic fields and are caused by accelerating electric charges. The two oscillating fields are at right angles to each other and to the direction of the wave.

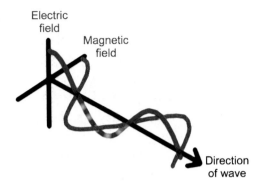

An electromagnetic wave (Credit: Peter Grego)

What information can we get from EM waves? As well as treating electromagnetic radiation as a wave, it was Einstein in 1905 who proposed that light and other EM radiation can be considered as individual photons; the higher the frequency of the EM wave, the higher the energy the photons carry.

Information from photons of electromagnetic radiation carries four types of information:

1. Spatial information
2. Temporal information
3. Spectral information
4. Polarization information.

Taking each category in turn:

1. *Spatial information.* Any telescope will image an astronomical object onto a two-dimensional plane where a detector, for example a CCD camera, will record the image. Apart from stars (which appear as point-like objects), planets, nebulae, star clusters, and galaxies will show themselves as extended objects, and the resolution will be limited by the size of the telescope's objective, the seeing conditions (if the observation is made through Earth's atmosphere), and the quality of the telescope's optics.

2. *Temporal information.* The variation of brightness of an astronomical object over time can be monitored – from incredibly rapidly spinning pulsars that fluctuate in brightness in the space of mere milliseconds to long-period variable stars whose changes in brightness

can take many years. Some of the latest astronomical research involves the detection of exoplanets – planets in orbit around distant stars – that have been discovered using the phenomenon of minute but measurable drops in brightness of a star when a planet moves in front of it, obscuring a small portion of the star's light.

3. *Spectral information.* By splitting the incoming EM radiation into its different wavelengths and observing how its intensity varies with wavelength, an enormous amount of information can be gleaned. Observations of a star's spectral lines and bands gives us an abundance of information, most obviously in being able to determine its surface temperature and the density of its atmosphere. Radial velocity of a star towards or away from the observer can be deduced through the Doppler effect; additionally, the Zeeman effect can show the strength of strong magnetic fields around a star, and the sinusoidal movement of spectral lines can indicate the presence of binary star systems.

4. *Polarization information.* You may have used Polaroid sunglasses in order to reduce the glare of reflected light from water; light reflected from the water is horizontally polarized, and this light will not be transmitted by the vertically aligned Polaroid glass. When light scatters off interstellar dust grains the light will be polarized to an extent, which can be measured with a polarimeter. Reflection nebulae – where a recently formed star is embedded in dust – will produce beautiful symmetric patterns of polarization due to scattering off dust particles. Polarization can also be found in high-energy sources such as active galactic nuclei (AGNs), where the mechanism is synchrotron radiation (relativistic charged particles that spiral in magnetic fields). Also, the cosmic microwave background (CMB) radiation from the Big Bang is now being studied using polarimetry. It is very weakly polarized, but this data is providing limits to models of an inflationary Universe.

Each day the amount of information being generated by all the space-borne satellites and ground-based observatories is enormous, currently amounting to 100 terabytes of information, or the equivalent of 100 million books! During the course of a year this adds up to more than 30 petabytes (1 petabyte = 1 billion books) of observational information. Over the next decade we will probably

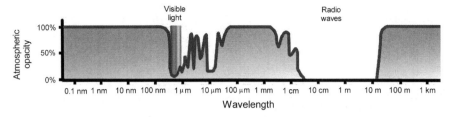

Earth's atmosphere has a range of opacities to the electromagnetic spectrum (Credit: Peter Grego)

see these figures increase by more than ten times – indeed, using Moore's law they might actually double every 2 years.

Only radio and visible light waves are able to fully penetrate the Earth's atmosphere, which means that ground-based observatories must concentrate upon examining these wavelengths. It stands to reason that the largest optical telescopes are best placed on tops of mountains under dry, dark skies in order to maximize seeing conditions and get above as much water vapor as possible. Examples of world-class high altitude observatories include those on Mauna Kea, Hawaii (4,200 m), La Palma on the Canary Islands (2,400 m), and Cerro Paranal, Chile (2,600 m), each of which is high enough to enable observations in near infrared wavelengths.

No matter what part of the EM spectrum a telescope is designed to observe it must gather as many photons as possible by using as large a collecting area that can be achieved. Thus the drive for telescopes of ever-increasing size has enabled astronomers to see increasingly fainter objects and in increasingly greater detail. A telescope's resolving power – its ability to see fine detail – is determined by several factors, namely the wave nature of EM radiation, the atmospheric seeing conditions (for optical astronomy in particular), and the quality of the telescope's optics.

Most astronomical objects are small and rather dim; the larger the telescope aperture the greater its light-gathering power and resolution, enabling finer, fainter detail to be seen. Optical telescopes have steadily increased in size over the last 400 years, and with this have come dramatic improvements in optical systems, eyepieces, and mounts. In four centuries the maximum size of optical telescopes has grown exponentially by 1.5% per year, from Galileo's 30-mm telescope in 1609 to a staggering 12 m in 2010. If this trend continues – and there is every reason to suppose that it will – we could predict that by the year 2100 there will exist a ground-based optical telescope whose main mirror will be 1,000 m in diameter.

Radio Astronomy

In 1931 Karl Jansky, working at Bell Laboratories, constructed a large rotating antenna in order to investigate the sources of static that might interfere with radio voice transmissions for a transatlantic radio telephone service. Initially thought to be of terrestrial origin, Jansky's "merry-go-round" identified a faint but steady radio "hiss" coming from the sky. After following the signal over a period of months, the loudest point of the emission repeated in a cycle of 23 h and 56 min, matching Earth's rotation period in space. Within a year of setting up his antenna, Jansky had determined that the signal was coming from the Milky Way and was strongest in the direction of Sagittarius, where the center of the Galaxy lay. This startling fact was confirmed in 1937 by Grote Reber's own radio observations made from an antenna built in his own backyard.

Although radio astronomy had been born, the Great Depression and then the Second World War prevented much progress being made on the astronomical front. One important wartime discovery was made in February 1942 by J. S. Hey working at the Army Operations Research Group in the UK. While investigating the cause of occasional strong interference to signals received by newly invented radar systems, Hey found that it was not disruption by German counter-weaponry but by bursts of intense radio emission from the Sun itself. Once the hostilities had ended, radio astronomy took off in a big way.

Ground-based radio telescopes operate in the wavelength range of between about 3 cm to 30 m. The iconic Lovell Telescope of Jodrell Bank Observatory in Cheshire, England, is the archetypical "big dish." Completed back in 1957 under the direction of pioneering radio astronomer Bernard Lovell, the 76-m diameter, fully-steerable dish was the largest of its type until 1971. The telescope was operational just in time to detect Sputnik-1's booster rocket in October 1957; another Cold War scoop came in February 1966 when images beamed back from the Moon by Soviet probe

Luna 9 were first revealed to the world by Bernard Lovell using an ordinary newspaper teleprinting machine to convert the probe's signals into pictures of the lunar surface. The Lovell Telescope's long and distinguished history also involves the Search for Extra-terrestrial Intelligence (SETI) programs, radar measurements of the distance to the Moon and Venus, and deep-space studies including observations of masers, pulsars, and quasars, in addition to the detection of intergalactic gravitational lensing.

Since 2000 the title of world's largest fully-steerable radio dish has been held by the 100 × 110-m diameter Green Bank Telescope in West Virginia in the United States. The largest non-steerable radio telescope is the Arecibo Radio Observatory, nestled within a natural crater on a hillside in Puerto Rico; 305 m in diameter it operates between 50 MHz up to 10 GHz.

Remarkably large though these individual radio telescopes are, a technique known as aperture synthesis through interferometry enables modern radio telescopes to be linked, making them equivalent to a telescope that is as large as the maximum baseline distance between the antennae in the network (interferometry has also been accomplished very successfully in the optical and infrared regions). The EM waves are combined from two or more collectors and made to produce interference patterns that are analyzed by computer. By doing so, far greater resolution is obtained, and more detailed observations of celestial objects can be made.

In order for a radio telescope operating at a wavelength of 50 cm to match the resolution of an amateur-sized 120-mm optical telescope operating at 500 nm (nanometers), a radio dish with a diameter of 120 km would be required; aperture synthesis using radio interferometry can deliver such a vast collecting area. Very Long Baseline Interferometry (VLBI) is performed by e-Merlin network in the UK; with its seven dishes (including the Lovell Telescope) and a maximum baseline of 217 km, e-Merlin has a resolution of 40 milliarcs – comparable to the visual resolution of the Hubble Space Telescope. An example of VLBI in the United States is the VLBA, consisting of ten radio dishes spanning North America from Hawaii to the Virgin Islands, with an 8,600 km baseline. Europe's EVN has 18 antennae and a maximum baseline of 8,200 km, which can be extended to 10,900 km in collaboration with the Arecibo Observatory in Puerto Rico.

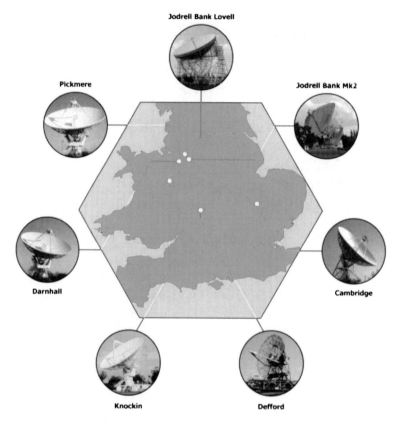

The impressive Merlin (Multi Element Radio-Link Interferometer Network) in the UK

New developments on the radio astronomy horizon include SKA (the Square Kilometer Array), which will have an area of one million square meters. This installation, possibly complete by 2020, will be the world's largest single radio telescope and cost approximately $1.6 billion. More than 15 countries are collaborating in the project, and by 2012 the site for this enormous telescope will have been chosen. It is likely to be located in the southern hemisphere, either Australia or South Africa. SKA will have an incredible sensitivity, and if any nearby exoplanets are broadcasting TV signals we will be able to tune in! It's amazing to think that within just 80 years of Grote Reber's pioneering 9-m telescope, SKA will represent an increase of 15,600 times in collecting area.

What Can You "See" with a Radio Telescope?

Stars like our Sun do emit some energy in radio wavelengths, but most radio emission comes principally from gas clouds (thermal emission and masers), pulsars, and radio galaxies (synchrotron emission). One very important radio spectral line is the 21.1 cm hydrogen line that was predicted by Hendrik van de Hulst in 1944 and detected in 1951. The 21.1 cm emission is due to neutral hydrogen (HI), which is found throughout our Galaxy. Some 10% of the observable mass of our Galaxy is in the form of gas clouds, in particular the giant molecular clouds (GMCs), which can contain 100,000 solar masses of hydrogen each and have typical temperatures in the range 8–60 K. GMCs, such as the Orion Nebula just 1,500 light years from us, are the places where star formation occurs.

Postulated by Lev Landau in 1932, pulsars are fast-spinning neutron stars that were first observed by Jocelyn Bell and colleagues led by Anthony Hewish in 1967. Since then astronomers have discovered more than 1,800 pulsars in our Galaxy. These objects allow astronomers to find out about matter at incredible densities – a thimble full of neutron star matter will have a mass of a billion tons.

Other strong radio sources are supernovae remnants. Examples include Cas A, marking the site of a star in Cassiopeia that was observed to explode some 340 years ago (11,000 light years distant), and the very strong Sgr A (Sagittarius A), which resides at the very center of our Galaxy. There is evidence that the source of energy for Sgr A is a supermassive black hole up to four million solar masses; one star orbiting this object has a velocity of 5,000 km/s, 1.6% the speed of light. Also observed in the radio region are rare radio galaxies, which number around one in a million of optical galaxies. Radio galaxies are very powerful emitters of radio waves, emitting up to 250 billion Suns' worth of energy at just radio wavelengths!

Eavesdropping on ET?

Most experts think that any intelligent civilization wanting to communicate to other parts of its home galaxy would use radio transmissions. Interstellar dust grains absorb or scatter light in the Galactic plane, so this means we can only peer through the "fog" of interstellar gas and dust out to 3,000 light years, while many objects give off infrared radiation, making this region of the EM spectrum very bright. Radio waves can easily pass through clouds of gas and dust. To maximize the chance of detection the transmissions would probably be over a narrow bandwidth, but the number of possible frequencies to broadcast would be enormous.

What about choosing the right star/solar system to tune into? It is thought that stars in spectral class range F5 to K5 will offer the best chance for intelligent life to evolve. Within 1,000 light years of Earth there are more than ten million Sun-type stars, and some of these solar systems may harbor life. So, in which direction should we point our radio telescopes? How many of these solar-type stars within 1,000 light years of Earth will have planets that are habitable? How many will have life on them, and how many will have intelligence and the technology to broadcast?

One way of calculating the odds is to use the famous Drake equation first formulated by Frank Drake in 1961.

$$N = N^* . f_p . n_e . f_l . f_i . f_c L / T_{gal}$$

where N = number of intelligent civilization who wish to and can communicate now N^* = number of stars in our Galaxy (to be between 100 and 200 billion), f_p = fraction of stars that have planets, n_e = number of planets that can support life per solar system, f_l = fraction of planets that will develop life, f_i = fraction of life-bearing planets that develop intelligent life, f_c = fraction of intelligence that can and will communicate, L/T_{gal} = fraction of the lifetime of our Galaxy in which the intelligent civilization survives.

The Drake equation applied to our Galaxy

N^* (in billions)	f_p	n_e	f_l	f_i	f_c	L/T_{gal}	N (number of civiliza-tions)
100	0.1	0.1	0.2	0.001	0.05	0.0000005	0.005
150	0.4	0.4	0.5	0.1	0.1	0.000004	480
200	0.6	0.7	0.9	0.3	0.2	0.00001	45,360

The table above takes a range of values to denote low, medium, and high values of the various factors. Thus, taking the product of f_p and n_e, it seems likely that between 1 and 42 in a hundred will have an Earth-like planet that could support some type of life. Multiplying out all the factors, then, the number of possible intelligent civilizations at the present time could be range between 1 and 40,000 within our Galaxy. Let us assume that just one is located within 3,000 light years of Earth.

Assuming a planet has intelligent life, how long will an intelligent civilization last, and for how long will it be broadcasting to the Universe? Life on Earth seems to have started roughly 800 million years after its birth and has lasted some 3.8 billion years. That big-brained bipedal primate homo-sapiens sapiens has been around for more than 50,000 years, but we have only been broadcasting for less than a century. If we take the probability that we are looking at a planet with intelligent life and not just life, a conservative estimate of the rough probability of listening in at the right time would be 100,000 years/3,800,000,000 years, or 0.003%. Sadly, 100,000 years of civilization appears very much an overestimate when we observe the human race in its present state. Thus the chances of "listening in" on an intelligence around a solar-type star are pretty slim.

Of course, we can reduce these truly astronomical odds by searching through tens of thousands of channels simultaneously and looking at many thousands of stars over a few decades. Since the discovery of the first exoplanet in 1992 we have located more than 490 planets (as of August 2010), including one Earth-sized planet in a habitable zone, Gliese 581d, comparatively nearby at only 20.5 light years from Earth. By targeting these solar systems the chances of discovering communication from ET might be improved, but still must be less than one chance in 100 billion. Still, if we don't try we most definitely won't find an alien civilization out there.

The first radio searches were started nearly 50 years ago in 1960 (Project Ozma). If you have a computer and Internet connection and would like to participate in a scientific project searching for signs of alien civilization, point your browser to the SETI (search for Extraterrestrial Intelligence) website run by the University of California, Berkley: http://setiathome.berkeley.edu.

An important radio telescope array called ATA (Allen Telescope Array) is being built at Hat Creek in California by the University of California, Berkley, and the SETI Institute. The first phase was completed in 2007 with 42 antennae, but there are planned to be some 350 6.1-m-diameter antennae by about 2012. These will simultaneously make astronomical observations and do SETI (search for extraterrestrial intelligence) searches, monitoring a million stars. It will be the largest radio telescope devoted to searching for extraterrestrial intelligence. However, it is versatile and can have several users simultaneously doing SETI searches and astrophysics observations.

Microwave Astronomy

In 1964, using the Bell Labs' 6-m horn antenna in New Jersey, U.S. scientists Arno Penzias and Robert Wilson discovered a low level, diffuse "hiss" permeating their signals. Puzzled by this, they set out to eliminate all possible mechanical and instrumental causes – including pigeon droppings in the antenna – yet the signal stubbornly remained. It became clear that the radiation was not being emitted by terrestrial sources, nor did it originate from the Sun or the Galaxy. Eventually it was concluded that the signal was nothing less than a relic of the hot primordial fireball, dating to around 380,000 years after the creation of the Universe in the Big Bang 13.7 billion years ago. Originally a temperature of 3,000 K, the expansion of the Universe has redshifted the radiation into microwave wavelengths, where it permeates the Universe as the cosmic microwave background (CMB) at a temperature of just 3° above absolute zero.

Just 25 years after the discovery of the CMB, the Cosmic Background Explorer satellite detected minute fluctuations in its intensity. Confirmed by other more sensitive instruments, notably the Wilkinson Microwave Anisotropy Probe (WMAP), these slight temperature variations within the CMB are, in effect, the first signs of the large-scale changes in the once-smooth fabric of space that were to lead to the formation of galaxies.

ESA's Planck mission, a 1.5-m space telescope launched in 2009, is taking a detailed look at the CMB from its stable Lagrangian vantage point some 1.5 million km from Earth. Planck's instruments provide ten times better sensitivity than previous space probes and are measuring anisotropies at the two parts in a million level with

The CMB temperature anisotropy observed by WMAP (Photo courtesy of NASA/WMAP Science Team)

a resolution down to 10 arcmin over the whole sky. In 2012 the European Southern Observatory (ESO) project ALMA (the Atacama Large Millimeter Array) in Chile will be operating between wavelengths of 1 cm down to 0.3 mm at an altitude of 5,000 m in Chile. ALMA, composed of 64 12-m antennae with a 10-km baseline, will have an impressive resolution of 0.01 arcsec, which is five times better than the Hubble Space Telescope operating in the visible part of the spectrum. It will have the ability to see evolving galaxies at the earliest epoch after the Big Bang, measure physical conditions of star-forming regions in galaxies throughout the Universe, see very fine details in star-forming regions, and observe the kinematics of active galactic nuclei down to spatial resolutions of 10 parsec. It will also see how planets form around pre-main sequence stars.

Infrared Astronomy

Infrared radiation was first detected by astronomer William Herschel in 1800 after observing the heating effect of different colors. By simply measuring temperature using a thermometer while placed in the spectrum of light from a prism, he discovered an increase in temperature beyond the red end of the spectrum at wavelengths that were invisible to the naked eye.

Celestial objects that emit in infrared wavelengths are in the temperature range of between 20 and 500 K, so infrared observations enable astronomers to study dust and gas in nebulae where star formation is occurring, and of course the study of exoplanets, many of which have surface temperatures in this range (the average surface temperature of Earth is 290 K). With a wavelength range of 0.7–300 μm, infrared radiation from space finds it difficult to penetrate Earth's atmosphere, as it is absorbed by water vapor; there are some windows in the near-infrared between about 0.7 and 11 μm, but to take advantage of this infrared telescopes need to be above as much water vapor as possible. Earth's atmosphere is opaque in the mid- to far infrared, so observations in these wavelengths need to be made by satellite observatories.

In 1983 IRAS (the Infrared Astronomical Satellite) successfully mapped the entire sky at infrared wavelengths (12, 25, 60, and 100 μm) and mapped more than 250,000 infrared sources.

Launched in May 2009, ESA's Herschel space telescope has established itself as a great infrared space observatory. Herschel (named after the great astronomer who discovered infrared radiation)

Artistic view of IRAS, an important infrared satellite that operated during the 1980s (Photo courtesy of ESA/NASA)

has a 3.5-m mirror and observes in mid- to far-infrared wavelengths. It is the first space telescope to observe in this waveband and is the largest primary mirror to have so far been launched. Like a number of other satellites, Herschel operates 1.5 million km from Earth at a stable Lagrangian point; the telescope is cooled to temperatures below 2 K using liquid helium. Herschel's mission is to detect the first galaxies formed after the cosmic dark age (just 500 million years after the Big Bang), to observe star formation, map the chemistry of our Galaxy, and to study the chemistry of the atmospheres of planets, satellites, and comets.

For more than 20 years the 2.4-m Hubble Space Telescope (HST) has successfully operated in Earth orbit, helped by the occasional servicing missions of shuttle astronauts, the final upgrades having been installed in May 2009. HST's successor, the James Webb Space Telescope (JWST), is to be a giant 6.5-m telescope – more than 120 times the collecting area of IRAS – operating in the near- to mid-infrared range (0.6–28 μm). Due for launch in June 2013, JWST will be placed at a Lagrange point 1.5 million km from Earth. Its design lifetime is between 5 and 10 years.

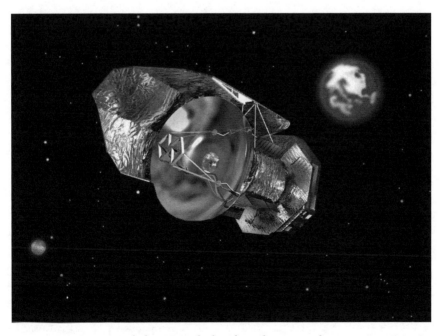

Artistic impression of the Herschel infrared space telescope in operation (Courtesy of ESA/Peter Grego)

Artistic impression of the James Webb Space Telescope in operation (Photo courtesy of ESA/NASA)

Optical Astronomy

Objects with a surface temperature of between about 2,900°C and 8,000°C will emit most of their radiation in the visible light region of the EM spectrum. Most of the ordinary (baryonic) matter in a galaxy is composed of stars; baryonic matter in our Galaxy is 90% stars, 9% gas, and 1% dust. The spectrum of light spans a tiny range of the EM waves from 380 to 750 nm. Of course, these are the wavelengths that our eyes can detect.

The Hubble Space Telescope (HST) was launched in 1990 into a low Earth orbit at 600 km altitude. After its final servicing in 2009 it is hoped that HST will keep actively observing until 2014 and possibly beyond. With a 2.4-m primary mirror (giving a resolution of 0.05 arcsec) the telescope has provided an unprecedented view of our Universe, both in visible light and parts of the infrared and ultraviolet. HST has refined our knowledge of the age of the Universe to 13.7 ± 0.1 billion years and has discovered evidence of dark energy and the acceleration of the Universe. In addition, HST has provided observations of protoplanetary disks, and nearer to home has kept a watchful eye on a wide range of activity within our Solar System. Prompting more than 6,000 scientific papers from its observations, HST still produces some 120 gigabytes of data every week.

An extended image known as the Hubble Ultra Deep Field has captured more than 10,000 galaxies in a field of 11.5 arcmin^2 (one 13 millionth of the whole sky). It follows that the total number of galaxies in our Universe could be 10,000 times 13 million, or at least 100 billion, each one with some 100 billion stars.

One optical space telescope that is expected to revolutionize the search for Earth-like planets is NASA's Kepler mission, launched in March 2009. It has a Schmidt telescope with a 1.4-m primary mirror, delivering a large field of view of 12°; every half hour a 95 megapixel camera performs transit photometry on 100,000 stars in the Lyra-Cygnus area of the sky. It is hoped that during its planned lifetime of nearly 4 years Kepler will find many exoplanets, including an estimated 500 plus Earth-like terrestrial planets.

Hopefully following the success of Kepler, NASA's TPF (Terrestrial Planet Finder) will consist of two space observatories – TPF-C

The Hubble Space Telescope following its final servicing mission in May 2009 (Photo courtesy of NASA)

Hubble Ultra Deep Field
Hubble Space Telescope • Advanced Camera for Surveys

NASA, ESA, S. Beckwith (STScI) and the HUDF Team   STScI-PRC04-07a

A magnificent ultra-deep field captured by the Hubble Space Telescope (Photo courtesy of NASA/STScI)

in 2014 and TPF-I in 2020. The pair will look for Earth-like planets within 300 light years, measuring size, temperature, distance from star, and the chemical composition of the planets' atmospheres. TPF will study the formation of planets from their beginnings in circumstellar disks around newly formed stars to detecting signs of life.

ESA's Darwin project – a search for Earth-like planets – is due to be launched about 2023. Darwin's design calls for four or five 3-m space telescopes that will survey the nearest thousand stars. Trying to image a planet close to a star in the visible spectrum requires discerning a contrast ratio of star to planet of a billion to one, but at cool infrared wavelengths (which the planet mostly emits) this ratio drops to just a million to one.

Still, the race is on to produce the next generation of very large ground-based optical telescopes, many of which will cost between $500 and $1,000 million.

The Giant Magellan Telescope (GMT) will consist of seven 8.4-m mirrors whose combined light-gathering ability will be equivalent to one giant mirror 24.5 m in diameter. GMT should be ready by 2018 and provide images ten times sharper than the Hubble Space Telescope. It has been designed to determine the large-scale properties of the Universe and the distribution and nature of its matter. One possible site for GMT is Las Campanas in northern Chile.

The TMT (30-m Telescope) is a collaboration between Caltech, the State University of California, and the Association of Canadian Universities for Research in Astronomy. The TMT may be built at one of two sites: Cerro Armazones, in Chile's Atacama Desert, or Mauna Kea on the Island of Hawaii. Possibly operational by 2018, TMT will consist of 492 1.4-m hexagonal mirror segments; it will be capable of directly imaging planets orbiting other stars, looking for the first stars and galaxies, and the epoch of reionization. It will have a wavelength range of 0.31–28 μm.

After originally commissioning a design study for a 60- to 100-m optical telescope (OWL: Overwhelmingly Large Telescope),

Artistic impression of the European Extremely Large Telescope (Photo courtesy of ESA)

the ESA is now concentrating on a smaller project called the E-ELT (the European Extremely Large Telescope).

E-ELT will have a 42-m main mirror made up of 984 segments that are actively controlled. Construction is to be started in 2012 – probably near the ESO at Paranal in Chile – and the telescope should be ready by about 2018. Once operational, the instrument will be able to detect the first galaxies formed after the Big Bang and image Earth-like exoplanets. E-ELT will seek to measure the initial mass function of stars at all epochs and examine host galaxies where gamma ray bursters have been observed. The telescope's magnitude limit will be magnitude +35 – five magnitudes fainter than the Hubble Space Telescope's limit. Proposed sites being looked at are either in northern Chile or the Canary Islands. The E-ELT is set to be the world's largest telescope, but for how many years? A 100-m telescope would have a magnitude limit of magnitude +38 in 10 h of imaging integration time. That's an astounding five trillion times fainter than can be seen by the human eye.

Ultraviolet Astronomy

Ultraviolet radiation was discovered in 1801 by Johann Ritter (1776–1810). He observed that when silver chloride was placed in the spectrum of light from the Sun after being dispersed by a prism, invisible radiation beyond violet darkened the silver chloride very rapidly; this radiation was called ultraviolet. With a wavelength range spanning 10–380 nm, nearly all ultraviolet radiation from space is blocked by the Earth's atmosphere. Some, of course, reaches ground level, notably UV-A (400–315 nm), which causes sunburn.

Very hot stars such as Harvard Class W and O (Wolf Rayet stars) and old stars that light up planetary nebulae where the surface temperature can reach 100,000 K give off most of their energy in ultraviolet wavelengths. The ultraviolet photons ionize the interstellar medium in which they are embedded, producing HII regions of ionized hydrogen – a notable example of such a region is around the Orion Nebula.

One of the most successful satellite observatories was the NASA/ESA International Ultraviolet Explore (IUE), which operated from 1978 to 1996. IUE secured more than 100,000 spectra, mostly of stars and galaxies, but also examined comets such as Halley in 1986 and Hyakutake in 1996. It observed the remarkable supernova in the Large Magellanic Cloud in 1987, the breakup and explosions caused by Comet Shoemaker Levy hitting Jupiter in 1994, and provided data for more than 3,500 published scientific papers.

NASA's GALEX (Galaxy Evolution Explorer) mission operated between 2003 and 2008, during which period it gathered an immense amount of data on over 500 million galaxies in the range 125–280 nm. This UV survey of nearly the entire sky is being used to research how galaxies evolve, both in structure and in chemical composition as stars fuse hydrogen into helium and beyond.

From 1999 to 2007 NASA's FUSE (Far Ultraviolet Spectroscopic Explorer) satellite observed in ultraviolet wavelengths between 90 and 120 nm, producing high resolution data on interstellar gas clouds. One of its main aims was to measure the amounts of deuterium (a heavy isotope of hydrogen with one proton and one neutron in its nucleus) present in the interstellar medium. Deuterium was produced in the first 3 minutes of the Big Bang and is destroyed in stellar cores, so its abundance gives astronomers important clues to the evolution of our Galaxy. FUSE also observed dust debris around newly formed stars such as Beta Pectoris, measured the abundance of hydrogen on Mars (showing that large amounts of water had escaped Mars in the past), and saw extremely hot gas surrounding our Galaxy and also very hot intergalactic gas (one to ten million K). Astronomers are attempting to explain the energy mechanisms responsible for heating this gas to such high temperatures.

X-Ray Astronomy

X-rays were discovered by Wilhelm Roentgen (1845–1923) as long ago as 1895, but it wasn't until June 1962 that astronomical objects were first observed in these wavelengths when an Aerobee 150 rocket carrying an X-ray detector discovered Sco X-1, the brightest X-ray source besides our Sun. Scorpius X-1 is a low-mass X-ray binary. These X-ray binary systems normally consist of a compact neutron star or black hole pulling off gas from an evolved star and forming an incredibly hot accretion disk.

An artistic impression of Cygnus X-3 (Credit: Peter Grego)

X-rays have wavelengths in the range of about 0.01–10 nm, and thermal sources that emit at these wavelengths have temperatures in the region 1–100 million K, so X-rays reveal some of the hottest objects in the Universe. Our Sun's corona – the pearly white sheen visible around the Sun during a total eclipse – strongly emits X-rays and has a temperature of between one and three million K.

Ordinary lenses absorb X-rays The pinhole camera was finally also applied in modern science. During the mid-twentieth century scientists discovered that it could be used to photograph X-ray radiation and gamma rays, which ordinary lenses absorb. As a result, the pinhole camera then found its way onto spacecraft and into space itself.

NASA launched its first X-ray satellite, Uhuru (SAS-1), in 1970, and it discovered around 400 celestial X-ray sources; later in that decade NASA's Einstein (High Energy Astrophysical Observatory 2) satellite boosted the number of known X-ray sources to 7,000. ESA's XMM Newton was launched in 1999, the most sensitive X-ray observatory launched to date and still going strong; one of its most important discoveries was a high energy iron fluorescence line in a class of active galaxies called Seyfert galaxies.

Sources of X-rays in our Galaxy mostly emanate from X-ray binary stars and supernova remnants, while extragalactic sources include active galaxies and quasars that contain supermassive black holes (those with at least a million solar masses), and extremely hot gas that lies between galaxies. This intergalactic gas can have temperatures in the range 10–100 million K, but the mechanisms responsible for heating the gas to such extreme temperatures is still being investigated. Scheduled for launch sometime between 2015 and 2025, ESA's next great X-ray mission is to be XEUS/IXO. About 200 times more sensitive than XMM-Newton, XEUS/IXO will have a 4.2-m diameter mirror, and it is hoped that its data will help determine how supermassive black holes formed and grew in the heart of most galaxies, as well as improving our understanding of how the large-scale structure of the Universe evolved and how the baryonic matter in the Universe became enriched.

Gamma Ray Astronomy

Gamma rays have the shortest wavelength and the highest energies with wavelengths less than 0.1 nm (100,000 eV, compared to the energy of light photons, which is in the range of 1.5–3 eV). The corresponding surface temperature of a body whose peak thermal radiation is 0.003 nm is 1 billion K, so these gamma rays are photons that represent the most energetic phenomena known in the Universe. Now, gamma rays by their very nature are so penetrative that it is extremely difficult or impossible to form an image, so most instruments are used as basic counters, meaning the photons' energy is measured, providing spectral information.

In 1972 ESA's SAS-2 satellite provided the first detailed look of the sky in gamma rays, but the mission only lasted just over 6 months. In 1973 the discovery of enigmatic gamma ray sources known as gamma ray bursters (GRBs) that had been detected by the Vela spy satellites (in the late 1960s) were announced. These sources last typically for a few seconds or minutes before disappearing. They are extreme events in very distant galaxies, probably involving the collisions of compact objects such as neutron stars or black holes, or produced during the formation of supermassive black holes. At least 30 GRBs are now detected each month.

Between 1991 and 2000 NASA's Compton Gamma-Ray Observatory successfully examined the gamma ray sky and discovered more than 2,000 gamma ray bursters and 66 examples of a new type of object that was termed a "blazar." Blazars are enigmatic objects, a sub-class of active galactic nuclei seen edge-on and thought to be accreting supermassive black holes. Compton also mapped the distribution of the radioactive isotope Aluminum 26, which has a short half-life of just 730,000 years; the isotope must be continually replenished by supernovae, releasing this element into the interstellar medium.

Very high energy gamma rays in the energy range of 100 billion to 10 trillion eV have been detected by the ground-based HESS (High Energy Stereoscopic System) in Namibia, which consists of four 13-m telescopes (operational since 2003). HESS uses the fact that extremely high energy gamma rays produce millions of secondary particles as they pass through kilometers of Earth's atmosphere, creating Cherenkov radiation in a narrow cone that can be detected on the ground. HESS has good angular resolution of a few arcminutes and has detected VHE gamma rays from supernova remnants such as RCW86, which may be the supernova seen back in AD 185.

ESA's INTEGRAL gamma ray observatory – the most sensitive gamma ray observatory so far – was launched in October 2002 and is currently a fully functioning success.

High energy cosmic rays (mostly protons) produce gamma rays by their interaction with nuclei in the interstellar medium. Gamma rays are also produced by radioactive decay, which give rise to spectral lines in the energy range of 0.1–2 MeV. These gamma ray observations are the best evidence of nucleosynthesis in supernovae. Finally, high energy gamma rays can be produced by interactions of charged particles in strong magnetic fields and also due to a process called Compton scattering.

NASA's Fermi Gamma-ray Space Telescope was launched into a low Earth orbit in 2008. Examining the sky in the range of 20 MeV to 300 GeV, Fermi has two important instruments: the Large Area Telescope (LAT) and the GBM, which will monitor GRBs. The LAT has a reasonable resolution of a few arcminutes at the highest energies down to 3° at 100 MeV. It has a 20° field of view and scans the whole sky in just 3 h. Fermi's 5-year mission is to understand how cosmic rays are accelerated to such enormous energies in AGNs, to determine the upper energy spectrum of

GRBs, and to search for evaporating mini black holes by detecting their Hawking radiation. So far Fermi has discovered a new type of pulsar in supernova remnant CTA1 that only emits in the gamma ray part of the electromagnetic spectrum.

Artistic impression of the gamma ray observatory INTEGRAL (Photo courtesy of ESA)

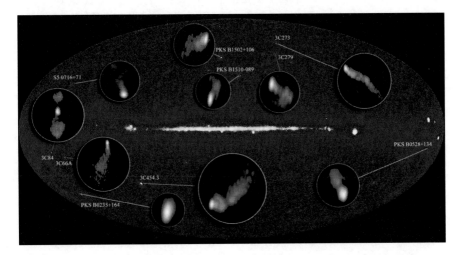

Radio jets of several active galaxies mapped by the Very Long Baseline Array (VLBA), inset into the Fermi Gamma Ray Space Telescope's map of the gamma ray sky (Photos courtesy of NASA/Fermi)

Neutrino Astronomy

Neutrinos are nearly massless particles that were postulated by Enrico Fermi in 1944 and first detected in 1956 by Frederick Reines and Clyde Cowan, Jr., at the Savannah River nuclear reactor. Cosmic ray neutrinos were detected in the early 1960's by Arnold Wolfendale et al. from the Kolar Gold mines, India. Approximately 65 billion neutrinos pass through every square centimeter of your body every second, yet out of this extraordinary number there might be just one neutrino interaction with a cell in your body in 70 years!

In the late 1960s Raymond Davis was the first to look for neutrinos coming from the Sun. Using a tank of 400,000 L of perchloroethylene (cleaning fluid), Davis was looking for the telltale signal of radioactive argon produced by an interaction of a solar neutrino with a chlorine nucleus. Over many years it was observed that only one third of the expected solar neutrinos were being detected. This discrepancy between calculated and observed flux was finally explained in 2001 by assuming that neutrinos had mass and that there are three "flavors" of neutrino: the electron, the muon, and the tau neutrino. Thus, between the point of production in the core of the Sun and reaching Earth some 8 min and 20 s later, over a distance of 150 million km, some of the solar neutrinos had changed flavor. A later neutrino experiment called Gallex was performed between 1991 and 1997 in the world's largest deep mine observatory in the Abruzzo region of Italy. Gallex used 30 tons of gallium and was located deep (beneath 2,400 m of rock) in the Gran Sasso mountain. The experiment successfully detected low energy solar neutrinos produced in the first proton-proton chain nuclear fusion reactions in the Sun's core, finding about 60% of the expected flux

In February 1987 a supernova in the Large Magellanic Cloud SN 1987A exploded and was visible with the naked eye for several weeks. It reached the third magnitude and was the first supernova visible without optical aid since Kepler's supernova of 1604 (indeed there have only been six supernovae seen with the naked eye in the last 2,000 years: AD 185, 1004, 1054, 1572, 1604, and 1987). On February 23, 1987, three separate neutrino experiments detected some 24 neutrinos in a short burst of 13 s from this supernova, 170,000 light years away. The core temperature of the SN 1987A will have briefly reached a maximum of 50 billion K, and in the rapid core collapse it gave off a neutrino burst that briefly equaled the neutrino flux from the Sun!

Gravitational Wave Astronomy

In his general theory of relativity in 1915, Einstein first predicted the existence of gravitational waves. Just as accelerated charges produce electromagnetic waves, then large masses orbiting each other will, through their accelerated motion, produce gravitational waves. These waves distort space-time and can induce a small change of length in objects many thousands of light years away. Traveling at the speed of light, the ripples carrying gravitational wave energy will pass through matter unhindered.

The search for gravitational waves started with Joseph Weber (University of Maryland) in the late 1960s, in which attempts were made to detect changes in length of a large, isolated cylindrical bar of aluminum using quartz strain gauges. The method failed, but it inspired many other workers in the field. Better methods for detecting the incredibly small changes in length induced by gravitational waves involved Michelson interferometers and high-powered, stable lasers.

At the moment there are several ground-based gravitational wave observatories, the largest of which is LIGO (Laser Interferometer Gravitational Wave Observatory) headed by Jay Marx and set up in 2002. LIGO consists of two observatories in the United States separated by over 3,000 km – one based in Hanford, Washington State, the other in Livingston, Louisiana. Both have two 4-km arms at right angles to each other and using laser interferometry can detect displacements of less than 1/1,000 the diameter of a proton over that length! Both observatories are linked, and the time difference between one detector detecting a gravitational wave and the second observatory could be up to 10 ms. When the gravitational waves impinge on the instruments they will produce minute but hopefully measurable distortions in space–time.

Other ground-based gravitational wave observatories include VIRGO in Cascina near Pisa, Italy, and GEO600, a German/ UK venture based in Hanover, Germany. Collaboration between

Aerial view of the LIGO interferometer at Hanford (Photo courtesy of LIGO Laboratory)

gravitational wave observatories such as VIRGO and LIGO and VIRGO and GEO600 will enable any event to be pinpointed in the sky and to reject spurious signals. It is hoped to detect gravitational waves from colliding neutron stars, supernovae, and even from the first few moments of the Big Bang. Very soon these observatories will be updated, giving rise to the Advanced LIGO, Advanced VIRGO, and so on. These instruments will be ten times more sensitive and are due to be completed in the next few years. At the time of this writing, however, no definite gravitational waves had been detected, and the sensitivity of instruments are such as to be able to detect two 1.4 solar mass neutron stars coalescing out to a distance of approximately 7 Mpc. By 2015 this distance will be 10 times larger and thus cover a volume of space 1,000 times greater. Up to 1,000 times as many events per year are expected to be detected.

Japan's LCGT (Large Cryogenic Gravitational wave Telescope) is to be built deep below the surface in the Kamioka mine by the University of Tokyo in the near future. It will consist of two 3-km arms with cryogenic mirrors operating around 20 K. LCGT will be sensitive enough to be able to detect two binary neutron stars coalescing out to 240 Mpc. It is hoped that the instrument could detect 2/3 of these events per year.

LISA (Laser Interferometer Space Antenna) is planned to be launched in 2018. Consisting of three satellites some five million km apart in an equilateral triangle, LISA is expected to be 100 times more sensitive than present day instruments. Even over the distance between the three satellites the change in length induced by a passing gravitational wave would be at most 1/100 the diameter of a hydrogen atom!

LISA and the planned Einstein telescope are third generation gravitational wave observatories and might be sensitive enough to detect primordial gravity waves from the earliest epoch of the Big Bang, some 380,000 years before the radiation we know as the cosmic microwave background. In the 2020s, the Gravitational Wave Cosmic Surveyor Explorer and the BBO (Big Bang Observer) will be the successors to LISA, but they currently remain on the drawing board.

Cosmic Ray Astronomy

Unlike all other sciences, astronomy is mainly an observational science. With the exception of only a few items that can be directly handled and studied, such as meteorites and particles from comets, scientists can never experiment with matter from outside our Solar System. In 1912, Victor Hess made a pioneering balloon flight during which he first discovered that there were particles with enormous energy hitting Earth's atmosphere. These particles have an energy range up to ten million times more energetic than the particles being accelerated in the Large Hadron Collider in CERN, Geneva.

These cosmic ray particles are mostly protons (hydrogen nuclei) with some heavier nuclei, e.g., helium, carbon, nitrogen, and oxygen. Due to the interaction of cosmic rays with the Galactic magnetic field, their arrival direction will not indicate their origin, unless one examines those cosmic rays with the highest energies. Cosmic rays with the very highest energy of 10^{20} eV are very rare, with the arrival of just one particle per square kilometer per century. How do we detect them? These high energy cosmic rays hit nitrogen and oxygen molecules in the atmosphere, producing huge showers of secondary particles that can be detected at ground level. The resultant extensive air shower can be then analyzed and the energy of the primary cosmic ray and its direction can be found. A 10^{19} eV cosmic ray will produce more than 10,000 million secondary particles that will spread out over an area at least 10 km^2.

One of the most successful cosmic ray observatories is the Pierre Auger Observatory at Malargue in Argentina, a collaboration of over 17 countries. It uses two techniques to detect the secondary particles: one utilizes the Cherenkov radiation produced by relativistic particles moving through a medium faster than light speed in that medium. These relativistic particles might pass through one of 1,600 water tanks, each with 3,000 L of pure water and the subsequent Cherenkov radiation is detected by photomultiplier tubes. The other technique utilizes the fluorescence produced by the secondary particles as they travel down through Earth's atmosphere and can be picked up using 24 giant fly-eye cameras. This technique was pioneered by the University of Utah Fly Eye experiment between 1981 and 1993, and later by the High Resolution Fly Eye in the late 1990s. In 1991 the Fly Eye instrument detected a particle whose energy was estimated to be a staggering 3.2×10^{20} eV!

A new Pierre Auger observatory is to be built in southeast Colorado that will build on the success of its predecessor in Argentina. Pierre Auger South has detectors covering an area of 3,000 km^2 (50 times the area of Manhattan), but Pierre Auger North is planned to cover an area seven times larger.

Finally, although great leaps have been made in our understanding of the Universe, let's not forget that we have mostly been looking at direct evidence of baryonic matter, which makes up just 4% of our Universe. There remains to be investigated the 21% dark matter and the even more mysterious 75% dark energy that is causing the expanding Universe to accelerate. So, here's a toast to the next 400 years of observational astronomy!

Appendix A

Best Astronomy Sites to Visit

United States

There are over 1,100 planetaria in the United States. Very good websites providing listings of numerous places to visit are:

http://www.astronomyclubs.com
http://www.go-astronomy.com
http://www.astronomy2009.us
http://www.amsky.com/calendar/events

Alabama
University of Alabama
Department of Physics and Astronomy
Tuscaloosa, AL 35487-0324
Tel: (205) 348 5050
http://www.astr.ua.edu

Alaska
Chugach Stargazer Society
707 A Street – Suite 201
Anchorage, AK 99587
Tel: (907) 269 7983
http://www.amsky.com

Arizona
Kitt Peak National Observatory Visitor Center
Tucson, AZ 85726-6732
Tel: (520) 318 8726
http://www.noao.edu/outreach

Lowell Observatory
1400 W Mars Hill Road
Flagstaff, AZ 86001
Tel: (928) 233 3211
http://www.lowell.edu

Arizona Science Center
600 East Washington Street
Phoenix, AZ 85004
Tel: (602) 716 2000
http://www.azscience.org

Arkansas
University of Central Arkansas
Department of Physics and Astronomy
201 Donaghey Avenue
Conway, AR 72035
Tel: (501) 450 5900
http://www.uca.edu

California
Palomar Observatory
35899 Canfield Road
Palomar Mountain, CA 92060-0200
Tel: (760) 742 2111
http://www.astro.caltech.edu/palomar

University of California, Los Angeles
Physics and Astronomy Building
430 Portola Plaza
Los Angeles, CA 90095-1547
http://www.astro.ucla.edu

Colorado
University of Colorado Boulder
Department of Astrophysical and Planetary Science
Boulder, CO 80309
http://cosmos.colorado.edu/sbo

Fiske Planetarium and Science Center
Regent Drive
Boulder, CO 80309
Tel: (303) 492 5002
http://fiske.colorado.edu/

Conneticut
The Discovery Museum
4450 Park Avenue
Bridgeport, CT 06604
Tel: (203) 372-3521
http://www.discoverymuseum.org

Delaware
Delaware Astronomical Society
http://delastro.org

Delmarva Stargazers
Tel: (302) 653 9445
http://www.delmarvastargazers.org

Florida
John F Kennedy Space Center
Kennedy Space Center, FL 32899
Tel: (321) 867 5000
http://www.ksc.nasa.gov

Kennedy Space Center Visitor Complex
Tel: (321) 449 4444
http://www.kennedyspacecenter.com

University of Florida
Department of Astronomy
211 Bryant Space Science Center
P.O. Box 112055
Gainesville, FL 32611-2055
Tel: (352) 392 2052
http://www.astro.ufl.edu/index.html

Georgia
Georgia State University
Department of Physics and Astronomy
29 Peachtree Center Avenue
Science Annex, Suite 400
Atlanta, GA 30303-4106
http://www.phy-astr.gsu.edu

Hawaii
Mauna Kea Observatory
Mauna Kea, Hawaii
Tel: (808) 974 4205
Tel: (808) 961 2180
http://www.ifa.hawaii.edu/mko

Idaho
Boise Astronomical Society
P.O. Box 7002
Boise, ID 83707
http://www.boiseastro.org

Illinois
University of Illinois at Urbana
Department of Astronomy
1002 W. Green Street
Urbana, IL 61801
Tel: (217) 333 3090
http://www.astro.uiuc.edu

Iowa
Cedar Amateur Astronomers, Inc.
P.O. Box 10786
Cedar Rapids, IA 52410
http://www.cedar-astronomers.org

Indiana
Indiana Astronomical Society, Inc.
http://www.iasindy.org

Kansas
The University of Kansas
Department of Physics and Astronomy
1082 Malott
1251 Wescoe Hall Dr.
Lawrence, KS 66045-7582
http://www.physics.ku.edu

Cosmosphere
1100 North Plum Street
Hutchinson, KS 67501–1499
http://www.cosmo.org

Kentucky
University of Kentucky
177 Chem.-Phys. Building
600 Rose Street
Lexington, KY 40506-0055
http://www.pa.uky.edu

Louisiana
Pontchartrain Astronomical Society
http://www.pasnola.org

Maine
Astronomical Society of Northern New England
http://www.asnne.org

Maryland
Goddard Space Flight Center
8800 Greenbelt Road
Greenbelt, MD 20771
http://www.nasa.gov/goddard

University of Maryland Observatory
http://www.astro.umd.edu/openhouse

Massachusetts
Amherst Astronomical Association
http://www.amastro.org

Michigan
Eastern Michigan University
Department of Physics and Astronomy
303 Strong Hall
Ypsilanti, MI 48197
Tel: (734) 487 3033
Sherzer Observatory
http://www.physics.emich.edu

Minnesota
Science Museum of Minnesota
120 W. Kellogg Boulevard
St Paul, MN 55102
http://www.smm.org

Minnesota Astronomical Society
P.O. Box 14931
Minneapolis, MN 55414
http://www.mnastro.org

Mississippi
Rainwater Astronomers Association
Rainwater Observatory
1 Fine Pl
French Camp, MS 39745
Tel: (662) 547 6377
http://www.rainwaterobservatory.org

Missouri
Astronomical Society of Kansas City
7241 Jarboe
Kansas City, MO 64114
http://www.askconline.org

Montana
Museum of the Rockies
600 West Kagy Boulevard
Bozeman, MT 59717
http://www.museumoftherockies.org

Nebraska
Fernbank Science Center
156 Heaton Park Drive N.E.
Atlanta, GA 30307
http://www.fernbank.edu

Nevada
Astronomical Society of Nevada
4275 Settler Drive
Reno, NV 89509
http://www.astronomynv.org

New Hampshire
New Hampshire Astronomical Society
http://www.nhastro.com

Dartmouth College
Department of Physics and Astronomy
6127 Wilder Laboratory
Hanover, NH 03755 3528
Tel: (603) 646 2854
http://www.dartmouth.edu/~physics/

New Jersey
New Jersey Astronomical Association
http://www.njaa.org

Amateur Astronomy Association of Princeton
P.O. Box 2017
Princeton, NJ 08543
http://www.princetonastronomy.org

New Mexico
The Albuquerque Astronomical Society
http://www.taas.org

New York
Amateur Astronomers Association of New York City
http://www.aaa.org

Amateur Observers' Society of New York, Inc.
http://www.aosny.org

North Carolina
Raleigh Astronomy Club
P.O. Box 10643
Raleigh, NC 27605
Tel: (919) 460 7900
http://www.raleighastro.org/

Morehead Planetarium
University of North Carolina
Chapel Hill, NC 27599
Tel: (919) 962 1236
http://www.moreheadplanetarium.org/

North Dakota
Dakota Astronomical Society
757 Munich Drive
Bismark, ND
Tel: (701) 223 2117

Ohio
Ritter Planetarium and Brooks Observatory
The University of Toledo
2801 W. Bancroft Street
Toledo, OH 43606-3390
http://www.utoledo.edu/as/rpbo

Perkins Observatory
P.O. Box 449
Delaware, OH 43015
Tel: (740) 363 1257
http://www.perkins-observatory.org

Wesleyan University
Department of Astronomy
Van Vleck Observatory
96 Foss Hill Drive

Middletown, CT 06459
Tel: (860) 685 2130
http://www.wesleyan.edu/astro

Oklahoma
Astronomy Club of Tulsa
P.O. Box 470611
Tulsa, OK 74147-0611
http://www.astrotulsa.com

Oregon
Rose City Astronomers
Tel: (503) 255 2016
http://www.rca-omsi.org

Pennsylvania
Amateur Astronomers Association of Pittsburgh
http://www.3ap.org

Rhode Island
Skyscapers, Inc.
Amateur Astronomical Society of Rhode Island
http://www.theskyscrapers.org

South Carolina
Midlands Astronomy Club, Inc.
http://www.midlandsastronomyclub.org

Tennessee
Sharpe Planetarium
Memphis Pink Palace Museum
3050 Central Avenue
Memphis, TN 38111
Tel: (901) 320 6320
http://www.memphismuseums.org

Barnard-Seyfert Astronomical Society
P.O. Box 150713
Nashville, TN 37315-0713
http://bsasnashville.com/

Texas
McDonald Observatory
Visitors Center
HC 75 Box 1337 – VC
Fort Davis, TX 79734
Tel: (432) 426 3640
http://www.mcdonaldobservatory.org

NASA Johnson Space Center
2101 Nasa Road 1
Houston, TX 77058
Tel: (281) 483 0123
http://www.nasa.gov/centers/johnson/home/index.html

Utah
Salt Lake Astronomical Society
http://slas.us/

Utah Skies
http://www.utahskies.org

Vermont
Vermont Astronomical Society
P.O. Box 782
Williston, VT 05495
Tel: (802) 879 4032
http://www.vtastro.org

Virginia
Science Museum of Virginia
2500 West Broad Street
Richmond, VA 23220–2057
Tel: (804) 864 1400
http://www.smv.org

Washington
Seattle Astronomical Society
http://www.seattleastro.org

Washington, DC
Smithsonian Institution
http://www.si.edu

National Air and Space Museum
6th Street and Independence Avenue SW
Washington, DC 20560
Tel: (202) 633 1000
http://www.nasm.si.edu

National Capital Astronomers
http://capitalastronomers.org

West Virginia
National Radio Astronomy Observatory
NRAO
P.O. Box 2
Green Bank, WV 24944
http://www.nrao.edu

Wisconsin
Madison Astronomical Society
Tel: (608) 233 7160
http://www.madisonastro.org

Universe in the Park
Department of Astronomy
UW-Madison
475 N. Charter St.
Madison, WI 53706
Tel: (608) 262-3071
http://www.astro.wisc.edu/uitp

Wyoming
University of Wyoming
Department Physics and Astronomy
Laramie, WY 82071
Tel: (307) 766 6150
http://faraday.uwyo.edu/

United Kingdom

England
Herschel House and Museum
19 New King Street
BATH
Avon
BA1 2BL
Tel: 01225 446 865

Thinktank at Millennium Point
Curzon Street
Birmingham Science Museum
Birmingham, B4 7XG
Tel: 0121 202 2222

Explore At-Bristol
Anchor Road
Harbourside
Bristol, BS1 5DB
Tel: 0845 345 1235
http://www.at-bristol.org.uk

Institute of Astronomy
University of Cambridge
Madingley Road
Cambridge, CB3 0HA
Carolin Crawford
Tel: 01223 337 510
http://www.ast.cam.ac.uk/public

Jodrell Bank Science Centre and Arboretum
Macclesfield
Cheshire, SK11 9DL
Tel: 01477 571 339
http://www.jb.man.ac.uk/viscen

Spaceport
Victoria Place
Seacombe Wallasey

Wirral, CH44 6QY
Tel: 0151 330 1333

Goonhilly
Helston
Cornwall, TR12 6LQ
Tel: 0800 679 593
http://www.goonhilly.bt.com

Norman Lockyer Observatory
Salcombe Hill
Sidmouth
Devon, EX10 0YQ
Tel: 01395 579 941
http://www.projects.ex.ac.uk/nlo

Taunton Space Walk
Bridgewater Canal
Devon
Tel: 01823 330 665
http://www.tauntondeane.gov.uk/Forms/Heritage/spacewalk.pdf

Worth Hill Observatory
Dorset
http://www.dstrange.freeserve.co.uk

Southend Planetarium
The Central Museum
Victoria Avenue
Southend on Sea
Essex, SS2 6EW
Tel: 01702 434 449
http://www.southendmuseums.co.uk

Bayfordbury Observatory
University of Hertfordshire
Hatfield, AL10 9AB
Tel: 01707 284 800
http://www.hertfordshire.ac.uk

Fort Victoria Country Park
Westhill Lane
Yarmouth
Isle of Wight, PO41 0RR
Tel: 01983 761555 or 0800 1958295
http://www.islandplanetarium.co.uk

Intech Science Centre and Planetarium
Telegraph Way
Morn Hill
Winchester
Hampshire, SO21 1HZ
Tel: 01962 863 791
http://www.intech-uk.com

Otford Solar System
Otford, Kent TN14
http://www.otford.org/solarsystem/

The Astronomy Centre
c/o North Midgelden Farm
Bacup Road
Todmorden
Lancashire, OL14 7HW
http://www.astronomycentre.org.uk

University of Central Lancashire
Centre for Astrophysics
Preston
Lancashire, PR1 2HE
01772 893 540
http://www.star.uclan.ac.uk

The National Space Centre
Exploration Drive
Leicester, LE4 5NS
Tel: 0116 261 0261
http://www.spacecentre.co.uk

Woolsthorpe Manor
23 Newton Way
Woolsthorpe-by-Colsterworth
nr Grantham
Lincolnshire, NG23 5NR
Tel: 01476 860 338

World Liverpool Museum
William Brown Street
Liverpool, L3 8EN
Tel: 0151 478 4393
http://www.liverpoolmuseums.org.uk

National Schools Observatory
Liverpool John Moores University
Twelve Quays House
Egerton Wharf
Birkenhead, CH41 1LD
Tel: 0151 231 2905
http://www.schoolsobservatory.org.uk

University of London Observatory
553 Watford Way
Mill Hill Park
London, NW7 2QS
Tel: 020 8959 0421
http://www.ulo.ucl.ac.uk

The Science Museum
Exhibition Road
London, SW7 2DD
Tel: 020 7942 4777
http://www.sciencemuseum.org.uk

National History Museum
Cromwell Road
London, SW7 5BD
Tel: 020 7942 5000
http://www.nhm.ac.uk

Royal Observatory Greenwich
Greenwich
London, SE10 9XJ
Tel: 020 8312 8565
http://www.nmm.ac.uk/astronomy

Norwich Astronomical Society
Seething Observatory
Toad Lane
Thwaite St Mary
Norfolk, NR35 2EQ
http://www.norwich.astronomicalsociety.org.uk

Keele University Observatory
Department of Physics and Astronomy
Staffordshire, ST5 5BG
http://www.astro.keele.ac.uk/~obs/

Newchapel Observatory
Off High St.
Newchapel
Stoke-on-Trent
Staffordshire, ST4 4PT
United Kingdom
http://www.alsager.com/tour/area/science.htm

Herstmonceux Observatory Science Centre
Hailsham
East Sussex, BN27 1RN
Tel: 01323 832 731
http://www.the-observatory.org/

The South Downs Planetarium and Science Centre
Sir Patrick Moore Building
Kingsham Farm
Kingsham Road
Chichester
West Sussex, PO19 8RP
Tel: 01243 774400
http://www.southdowns.org.uk/sdpt

The Wynyard Planetarium and Observatory
Thorpe Thewles
Stockton-on-Tees, TS21 3JG
Tel: 01740 630 544
http://www.wynyard-planetarium.net

The Planetarium
South Tyneside College
St. George's Avenue
Tyne and Wear, NE34 6ET
Tel: 0191 427 3589
http://www.stc.ac.uk/home/

Yorkshire Planetarium
Harewood House
Harewood
Leeds, LS17 9LG
Tel: 0113 218 1000
http://www.yorkshireplanetarium.co.uk

Northern Ireland and Eire

Armagh Planetarium
College Hill
Armagh, BT61 9DB
Tel: 028 3752 3689
http://www.armaghplanet.com

W5 at Odyseey
2 Queens Quay
Belfast, BT3 9QQ
Antrim
Northern Ireland
Tel: 028 9046 7700
http://www.w5online.co.uk

Birr Castle
BIRR
Co. Offaly
Ireland

Tel: 00 353 57 91 20336
http://www.birrcastle.com

Scotland

Aberdeen College Planetarium
School of Science and technology
Gallowell
Aberdeen, AB9 1DN
Tel: 01224 612 323
http://www.abcol.ac.uk

Mills Observatory
Glamis Road
Balgay Park
Dundee, DD2 2UB
Tel: 01382 435 967
http://www.dundeecity.gov.uk/mills

Dynamic Earth
Holyrood Road
Edinburgh, EH8 8AS
Tel: 0131 550 7800
http://www.dynamicearth.co.uk

Royal Observatory Edinburgh Visitor Centre
Blackford Hill
Edinburgh, EH9 3LU
Tel: 0131 668 8404
http://www.roe.ac.uk

Glasgow Science Centre
50 Pacific Quay
Glasgow, G51 1EA
Tel: 0871 540 1000
http://www.glasgowsciencecentre.org

Glasgow University Observatory
Acre Road
Mary Hill

Glasgow
Tel: 0141 330 4152
http://www.astro.gla.ac.uk/outreach

St. Andrews University Observatory
School of Physics and Astronomy
North Haugh
St. Andrews, KY16 9SS
Tel: 01334 463 103
http://www.st-andrews.ac.uk

Wales

University of Cardiff Observatory
School of Physics and Astronomy
Cardiff University
The Parade
Cardiff, CF24 3AA
http://www.astro.cardiff.ac.uk/observatory

Techniquest
Stuart Street
Cardiff Bay
Cardiff, CF10 5BW
Tel: 02920 475 475
http://www.techniquest.org

Techniquest is based in Cardiff but has four sites in Wales:

Spaceguard Centre
http://www.spaceguarduk.com

Llansahay Lane

Knighton

Powys
LD7 1LW
Tel: 01547 520 247

Appendix B

Useful Websites

The Astronomical League (US)
http://www.astroleague.org/

Society of Popular Astronomy (UK)
http://www.popastro.com

The Federation of Astronomical Societies website (UK)
http://www.fedastro.org.uk

American Astronomical Society
http://www.aas.org

Astronomical Society of the Pacific
http://www.astrosociety.org

Astronomy Now (UK astronomy magazine)
http://www.astronomynow.com

Bradford Robotic Telescope (UK)
http://www.telescope.org/

British National Space Centre (UK)
http://www.bnsc.gov.uk/

The Extrasolar Planet Encyclopaedia
http://www.exoplanet.eu

European Space Agency (ESA)
http://www.esa.int

European Space Organisation
http://www.eso.org

Faulkes Telescope (UK)
http://faulkes-telescope.com/

The Galileo Project: The Life and Work of Galileo Galilei (1564–1642)
http://galileo.rice.edu/

International Year of Astronomy 2009
http://www.astronomy2009.org
http://astronomy2009.us/
http://www.astronomy2009.co.uk

Liverpool Robotic Telescope (UK)
http://telescope.livjm.ac.uk/

National Aeronautical Space Agency (NASA)
http://www.nasa.gov

Royal Astronomical Society (UK)
http://www.ras.org.uk

Science and Technology Facilities Council STFC (UK)
http://www.scitech.ac.uk

Sky & Telescope
http://www.skyandtelescope.com
http://www.space.com
http://www.seds.org

Space Telescope Science Institute
http://www.stsci.edu

Stellarium Planetarium Software
http://www.stellarium.org

Appendix C

Mathematics Used in This Book

When making measurements astronomers use the SI Unit system on the whole, but many units are not standard SI, such as the astronomical unit (A.U.), the light year (l.y.), and the parsec (pc) for distance or arcsecond and arcminute for angle. Standard form is used to represent the enormously large or the incredibly small. It is the usual shorthand way of representing numbers using powers of ten. Any number can be represented by a number between 1 and 10 multiplied by a power of ten.

If we take powers of ten:

$10^2 = 10 \times 10 = 100$, while $10^3 =$ ten cubed or $10 \times 10 \times 10 = 1,000$. Thus we can write a million $(1,000,000) = 10^6$, a billion $(1,000,000,000) = 10^9$ and a trillion which is a million million $(1,000,000,000,000)$ is 10^{12}.

If the core temperature of the Sun is 14,000,000 K this is equivalent to

$1.4 \times 10,000,000$ K OR 1.4×10^7 K.

Take the average distance between Earth and the Sun as 149,500,000 km, which in standard form is
1.495×10^8 km and as 1 km $= 1,000$ m this distance in meters is
1.495×10^{11} m. When you multiply powers of ten you simply add the exponent

10^3 times $10^8 = 10^3 + 10^8 = 10^{11}$

279

Similarly, when dividing powers of ten, one subtracts the exponent, e.g.,

$$10^9/10^4=10^{9-4}=10^{11}$$

Examples of large numbers in astronomy:

1. If you assume a finite observable Universe then the number of stars given 100 billion stars in a galaxy on average and 100 billion galaxies is

$100\times10^9\times100\times10^9=10^{2+9+2+9}=10^{22}$. Taking a star like the Sun the total mass in the form of stars is 2×10^{30} kg$\times10^{22}=2\times10^{52}$ kg. One gram of hydrogen, the most common element found in stars, contains approximately 6×10^{23} atoms each with 1 proton and 1 electron so the total number of protons will be:

2×10^{52} kg$\times1{,}000\times6\times10^{23}\sim1.2\times10^{78}$!

2. The number of seconds since the Big Bang, assuming it took place 13.7 billions years ago, is a paltry $13.7\times10^9\times3.16\times10^7=4.32\times10^{17}$ s. The size of the visible universe in terms of the diameter of a proton is:

$10^{26}/10^{-15}=10^{41}$. The number called by mathematicians the Googol is much larger than any of these examples: 10^{100}! It is 1 with a hundred zeros.

Prefixes Used with SI Units

T	tera	10^{12}
G	giga	10^9
M	mega	10^6
k	kilo	10^3
c	centi	10^{-2}
m	milli	10^{-3}
m	micro	10^{-6}
n	nano	10^{-9}
p	pico	10^{-12}

One can put any prefix in front of an SI unit (e.g. mm is a millimeter). A micro kelvin or one millionth of a kelvin would be μK. The strength of magnetic fields in astronomy is often given in

Gauss instead of the SI unit, the Tesla. 1 Gauss = 10^{-4} Tesla. Thus a neutron star with a magnetic field strength of 10^{12} Gauss would be 100 MT. At the other end of the scale the galactic magnetic field is approx 1 μ Gauss or 0.1 nT.

Units of Angle

There are 360° in a full circle. The Moon and Sun both subtend an angle of approximately 0.5°. There are 60 arcminutes in 1°, so these two objects subtend an angle of 30 arcminutes.

Smaller angles use the unit known as the arc second where 60 arcseconds = 1 arcminute.

Thus 1° = 60 arcminutes (60′) = 3,600 arcseconds (3,600″).

To give you an idea on how small these angles are:

1 arcminute is equivalent to a Nickel coin (21 mm diameter) at a distance of 73 m, and 1 arcsecond would be the same coin at 4.4 km (nearly 3 miles).

Smaller angles such as the mas or milli arcsecond can viewed as a British 5 p coin (18 mm in diameter) at a distance of 3,125 km (or nearly 2,000 miles). The micro arcsecond μ as would mean the coin would have to be placed a thousand times further away or 3,125,000 km – 8 times further away than the Moon.

Glossary

Absolute zero   The coldest possible temperature in the Universe, which is 0 Kelvin (0 K) or –273.15°C. It is the temperature at which all atoms cease to move.

Acceleration   The rate of change of velocity. It is measured in m/s^2. The approximate value for the acceleration due to gravity at Earth's surface is 10 m/s^2, which means an increase in velocity of 10 m/s every second for a freely falling body ignoring air resistance.

Active galactic nuclei (AGN)   A collective name for very strong sources from active galaxies such as quasars, Seyfert galaxies, BL Lacertae objects, etc. These compact regions of active galaxies are thought to be powered by accretion onto supermassive black holes (more than a million solar masses).

Albedo   A measure of how reflective a body is, which is the percentage of incident light that is reflected back into space. The Earth has an albedo of 39% while the Moon's albedo is only 7%, similar to that of coal. Venus has a very high albedo of 65%, due to its cloudy atmosphere.

Altitude   The angle of an object above the observer's horizon. An object on the horizon has an altitude of 0°, while at the zenith its altitude is 90°.

Aperture   The diameter of a telescope's objective lens or primary mirror usually measured in centimeters or meters. At present (2010) the largest optical telescope is 11.8 m.

Arcminute   A measure of angle equal to 1/60 of a degree.

Arcsecond   A measure of angle equal to 1/60 of an arcminute or 1/3,600 of a degree.

Astrolabe An ancient astronomical instrument that combines a planisphere with sights to enable one to compute astronomical problems and measure time. The instrument may have been invented by Hipparchus in the second century BCE.

Astronomical unit (A.U.) The unit of distance equal to the mean distance of Earth as it orbits the Sun (149,598,000 km).

Aurora Also known as the Northern or Southern Lights; they are phenomena seen at high latitudes close to the magnetic North and South poles and are due to energetic charged particles trapped in Earth's magnetic field exciting oxygen and nitrogen atoms high up in Earth's atmosphere (50–80 km).

Baryonic matter A class of subatomic matter that includes neutrons and protons. You, me, the hundreds of billions of stars and planets are all made of baryonic matter, but this makes up just 4% of the total mass of the Universe!

Black body A hypothetical body that absorbs all incident radiation and emits radiation that is a function of its temperature only. The Sun is approximately a black body!

Black hole A body whose gravitational field is so strong as to stop all electromagnetic radiation, including light, from escaping.

Celestial sphere An imaginary sphere centered on Earth on which the Sun, stars, planets, and the Moon were thought to be located.

Centripetal force Any force that makes a body move in a circular orbit. The centripetal force always acts at right angles to the velocity of a body and towards the center of the circle.

Cepheid A type of variable star named after the prototype δ Cephei, which pulsates with a period of 5.33 days. Cepheid variable stars have periods of between 1 and 100 days. Henrietta Leavitt in 1912 discovered that the luminosity of a Cepheid is linked to its period.

Cherenkov radiation Radiation produced when charged particles with mass move faster than the speed of light in that medium.

Conjunction The close position as seen in the sky of two or more astronomical objects.

Cosmic microwave background (CMB)   The leftover radiation from the Big Bang, which decoupled from the matter in the Universe after 380,000 years when the universe had cooled to 3,000 K. The radiation now has a temperature close 2.73 K and has been redshifted by the expansion of the Universe into the microwave region. It was first discovered by Arno Penzias and Robert Wilson in 1965.

Cosmic rays   These are particles that have been accelerated to enormous energy of between 1 MeV (low energy cosmic rays from our Sun) up to a staggering 100 million TeV – the highest known cosmic ray energy. They were discovered by Victor Hess in 1912.

Dark matter   A mysterious and unknown form of matter that does not emit radiation, but is observed through its gravitational effects. It makes up 21% of the Universe.

Dark energy   A form of energy that fills the Universe and is responsible for the acceleration of its expansion. It makes up 75% of the total mass content of the Universe.

Electromagnetic radiation   Radiation that can travel through the vacuum of space and consists of an oscillating electric field at right angles to an oscillating magnetic field. Light is a form of electromagnetic radiation.

Electromagnetic spectrum   There are seven named regions of the electromagnetic spectrum, which in order of wavelength from longest to shortest are: radio waves/microwaves/infrared/light/ ultraviolet/X-ray/gamma rays.

Electronvolt (eV)   A unit of energy used by physicists at the atomic level. A photon of light has a typical energy in the range 1.5 (red light) to 3 eV (violet light). $1 \text{ eV} = 1.60 \times 10^{-19}$ J.

Elongation   The apparent angular distance of an object from the Sun, measured between 0 to 180° east or west of the Sun. For example, the first quarter Moon has an eastern elongation of 90°; Venus has a maximum possible elongation of 47°.

Escape velocity   This is the velocity of an object that is required to leave the gravitational field of a large body such as a planet or a star. The escape velocity required to leave Earth is 11.2 km/s.

Exoplanet   A planet that orbits a star that is not our Sun. Since 1990 there have been more than 490 exoplanets discovered (as of August 2010).

Galaxy   A huge conglomeration of stars that are held together by mutual gravity and range in size from dwarf galaxies, containing a few million suns, to giant elliptical galaxies that are ten times the mass of our Milky Way Galaxy.

Galilean moons   The four satellites seen by Galileo in January 1610, which have been subsequently named Io, Europa, Ganymede, and Callisto.

Geocentric universe   The theory that was adopted by Ptolemy in his great work the *Almagest* that all bodies orbit Earth including the five planets, the Sun, the Moon, and all the stars.

Giant molecular cloud   Huge galactic cloud of mostly molecular hydrogen (H_2) up to 100,000 solar masses that are very cold ~10–60 K. Some parts of giant molecular clouds can be regions of star formation in a galaxy.

Gravitational wave   A ripple in space-time (predicted by the General Theory of Relativity) that travels at the speed of light and will affect the dimensions of a body only slightly when it passes through the body.

Hertzsprung-Russell diagram   A graph that plots the luminosity of stars (y axis) against their surface temperature (x axis), devised independently by two astronomers: Ejnar Hertsprung and Henry Russell.

Heliocentric universe   A view of our Solar System in which all the planets, including Earth, revolve around the Sun. It was the Polish astronomer Nicolaus Copernicus who expounded this idea in his great book *de Revolutionibis Orbium Coelestium*, published in 1543 on his deathbed.

Interferometry   The technique by which two or more telescopes can be combined such that their spatial resolution is equivalent to a telescope whose diameter would be equal to the maximum distance between the telescopes.

kelvin The SI unit of temperature. The link between temperature in degrees Celsius and kelvin is approximately t (in °C) = T (in K) + 273. So a room temperature of 20°C would be 293 K.

Lagrangian point There are two Lagrangian points in which small bodies may orbit about the Sun without being perturbed by a planet, and these are at 60° ahead and behind a planet in its orbit around the Sun.

Luminosity This is a measure of the power output of a star and can measured in watts (W). The Sun's luminosity is 4×10^{26} W, which is enough energy produced in 1 s to last the Earth's human population for a 100,000 years on Earth!

Light year The distance light or any other form of electromagnetic radiation travels in the vacuum of space in a year. The distance is about 9.5 trillion km, or 63,200 AU.

Magnitude A unit of brightness of an astronomical object which has a logarithmic scale such that a difference of 5 magnitudes represents a change of 100 times in light intensity. The Sun is the brightest object with magnitude –26.7, while the full Moon is –13. The dimmest stars that can be seen by the naked eye are magnitude +6.

Main sequence star Stars on the main sequence of the Hertzsprung-Russell diagram are ones that derive their energy from nuclear fusion of hydrogen into helium. Ninety percent of a star's lifetime is spent on the main sequence. The Sun is a main sequence star and is halfway through its lifetime of 9 billion years.

Maser Cosmic masers were first discovered in 1965 and are the astronomical counterpart to laboratory masers. MASER is an acronym for Microwave Amplification by Stimulated Emission of Radiation and works on the principle of the laser but in the microwave region.

Muon (μ) A muon is an elementary particle similar to the electron and with a negative charge, but it is unstable and only lasts a few millionths of a second. Cosmic rays produce huge numbers of muons and several pass through your head every second! They were discovered by Carl Anderson in 1936.

Mural quadrant   An ancient astronomical instrument that was used to measure the altitude of celestial objects that crossed the observer's meridian.

Neutrino   An uncharged lepton with a mass less than one ten millionth of the mass of an electron that only interacts very weakly with matter. Neutrinos were postulated by Enrico Fermi in 1944 and discovered in 1956 by Reines and Cowan. It is now known that there are three flavors of neutrino: electron neutrino, muon neutrino, and the tau neutrino.

Neutron star   At the end of a massive star's life (mass > 10 M☉) the star will explode as a supernova and the remnant core can form a fast-spinning neutron star composed mainly of neutrons and with a size approximately 10 km in diameter. The density of a neutron star is similar to that of an atomic nucleus, so 1 cm^3 of neutron star will have a mass of a billion tons!

Nova   Very old star (white dwarf) that accretes material onto its surface from a binary companion, which can lead to a runaway thermonuclear explosion producing up to a million fold increase in brightness for a few days.

Oort cloud   A large spherical cloud of material that may lie some 50,000 to 100,000 AU from the Sun and contain many billions of comets. The idea was first suggested by Jan H. Oort in 1950. It is believed that the Oort cloud is the source of long-period comets whose orbital period around the Sun may be measured in thousands of years.

Parallax   The apparent position of a star changes as Earth orbits the Sun, and astronomical parallax is the measure of this angle in arcseconds. The nearest star, Proxima Centauri, exhibits a parallax of 0.76 arcseconds.

Parsec   A unit of distance equivalent to a parallax of 1 arcsecond. The parsec, or pc, is equivalent to 3.26 light years or 206, 265 astronomical units.

Photon   A particle of electromagnetic radiation whose energy is specified by Plancks equation $E = hf$ where h = Plancks constant 6.63×10^{-34} Js and f = frequency of radiation in hertz (Hz).

Planet   A body that orbits a star such as the Sun that is large enough to have become spherical through its own gravity and to have cleared its neighborhood of objects close to its orbit.

Ptolemaic system   The Ptolemaic system was based on the geocentric theory as described in the famous book the *Almagest* written by Ptolemy in AD 150 and was adopted by the Roman Catholic Church.

Pulsar   Fast-spinning neutron star that can emit radiation that appears as short pulses if the beam of radiation is in our line of sight (like a lighthouse beam). The pulses are extremely regular and range from milliseconds to several seconds. There are more than 1,800 pulsars that have been observed in our Galaxy, the first discovered by Jocelyn Bell in 1967. The Crab Nebula which is the supernova remnant of the supenova seen in 1054 by the Chinese, has a 33-ms pulsar that has been observed at all wavelengths from radio to gamma rays.

Quadrant   An ancient instrument used to measure the altitude of celestial objects above the horizon; used up until the seventeenth century.

Resolution   This is a measure of the ability of an optical instrument to resolve, or show, fine detail such as two close stars as separate entities rather than one star. Resolution is dependent on the aperture of the optical instrument and the wavelength of the radiation.

Retrograde motion   The outer planets exhibit retrograde motion as seen from Earth when they move in the sky from east to west rather than the normal west to east motion. This is due to Earth, which has a faster speed than the outer planets, catching up with the outer planet and then pulling away as it orbits the Sun.

Roche limit   The strong tidal forces of a large planet can disrupt a smaller orbiting body such as a satellite if it orbits at a distance less than the Roche limit, which was devised by Édouard Roche in 1848. For two bodies of similar density the critical distance is less than about 2.45 times the radius of the larger body. Examples of bodies orbiting a large planet less than the Roche limit are the ring systems of Jupiter, Saturn, Uranus, and Neptune.

Solar mass   A unit of mass used by astronomers equivalent to 1.99×10^{30} kg. Stars range in mass from 0.06 of a solar mass to 100 solar masses. The largest galaxies are over a trillion solar masses.

Spectroscope   An instrument that splits up electromagnetic radiation into its different wavelengths and is used to analyze how the intensity of radiation varies with wavelength. With a spectroscope, the astronomer, Sir Norman Lockyer, discovered the element helium in the Sun in 1868 before it was found on Earth.

Spectrum   The splitting of electromagnetic waves into its various component wavelengths. In terms of light one is splitting white light into its component colors. A spectrum can be analyzed to reveal many properties, including surface temperature, radial velocity, composition, and magnetic field strength.

Star formation region   A region of space where stars have recently been formed or are being formed from gas and dust in a part of a giant molecular cloud. The Orion Molecular Cloud 1,500 light years from Earth is the nearest star formation region.

Supernova   A cataclysmic explosion at the end of a star's life, which can release as much energy per second as the output of a whole galaxy, i.e. 100 billion stars. There have been six observed supernovae by the naked eye in the last 2,000 years.

Synchrotron radiation   When charged particles traveling close to the speed of light are accelerated in magnetic fields, then synchrotron radiation is released. The radiation was first postulated by I.S. Shklovski in 1953.

Terminator   The line separating the illuminated and unilluminated hemispheres of a planet or satellite.

Thermal radiation   This is electromagnetic radiation due to a body's thermal energy and is often similar to black body radiation in its spectrum.

Thermodynamic temperature   Temperature measured in kelvin, formerly known as absolute temperature.

TLP   Transient lunar phenomena. A rarely observed, short-lived anomalous colored glow, flash, or obscuration of local surface detail, whose causes are poorly understood.

Universal Time (U.T.)   The standard measurement of time used by astronomers over the world. UT is the same as Greenwich Mean Time, and it differs from local time according to the observer's position on Earth and the time conventions adopted in that country.

Universe   Our Universe started in a Big Bang explosion in which matter and space expanded from a singularity some 13.7 billion years ago. Recent measurements show that the expansion of the Universe is accelerating due to a mysterious form of matter called dark energy.

Zenith   The point in the sky directly above the observer. It has an altitude of 90°.

Index

About the Authors

Peter Grego is an astronomy writer and editor. A regular watcher of the night skies since 1976, he observes from his home in St Dennis, Cornwall, UK, with a variety of instruments. Grego's primary observing interests are the Moon's topography and the bright planets, but he likes to "go deep" when there's no glare of the Moon to contend with.

Grego has directed the Lunar Section of Britain's Society for Popular Astronomy (SPA) since 1984 and is the Assistant Director of the Lunar Section of the British Astronomical Association (BAA). He edits and produces four astronomy publications – *Luna* (journal of the SPA Lunar Section), the SPA News Circulars, and *Popular Astronomy* magazine. He is also layout editor for the Newsletter of the Society for the History of Astronomy.

Grego is the author of numerous astronomy books, including: *Collision: Earth!* (Cassell, 1998), *Moon Observer's Guide* (Philip's/Firefly, 2004), *The Moon and How to Observe It* (Springer, 2005), *Need to Know? Stargazing* (Collins, 2005), *Need to Know? Universe* (Collins, 2006), *Solar System Observer's Guide* (Philip's/Firefly, 2005), *Venus and Mercury and How to Observe Them* (Springer, 2008); *Astronomical Cybersketching* (Springer, 2009); *The Great Big Book of Space* (QED, 2010), and others. He has given many talks to astronomical societies around the UK and has been featured on a number of radio and television broadcasts.

Grego maintains his own website at www.lunarobservers. com (which occasionally features live webcasts of the Moon and planets and other astronomical phenomena) and is webmaster for the BAA Lunar Section at www.baalunarsection.org.uk. He is a member of ALPO, SPA, SHA, and BAA and is a Fellow of the Royal Astronomical Society.

David Mannion has three degrees in astronomy and has worked as a teacher for 23 years in schools and colleges in the UK, Austria, and Turkey, and has also tutored for the Open University in Physics and Astronomy. He is a Fellow of the Royal Astronomical

Society, having been elected in 1984 and was a member of its Education Committee 2005–2010.

Dr. Mannion has given lectures on astronomy since 1980, run numerous School Astronomy Clubs, and was vice president and a founder member of the Association for Astronomy Education.

His other burning interest is weightlifting and participating in the last 6 years of the British Masters Weightlifting Competition he has won in his age group and weight category in 2007 and 2010! He wants to continue watching the stars and lifting weights for as long as he possibly can!

Printed by Publishers' Graphics LLC